Protection of Environment

Protection of Environment

Aditya Bhatnagar

Oxford Book Company

Jaipur ▪ New Delhi

Published By

Oxford Book Company

Regd. Off. "Bony Residency", Gate No. 2
Opp. Tilak Public School, Vishweswariya Nagar,
Gopalpura Road, Jaipur - 302018, India
Ph: 0141 - 2761280, Telefax: 2761381
email - oxfordbookjpr@gmail.com
website - oxfordbookcompany.com

Delhi Office: 102, 1st Floor, Satyam House,
4327/3, Ansari Road, Darya Ganj,
New Delhi - 110002, India
Ph: 011-45652440

Published in India, 2025

ISBN : 9789350300015

Price : ₹ 2500/-

Typeset by Shivangi Computers, Jaipur

Printed and bound in India by Thomson Press Ltd.,
New Delhi

Preface

Protection of Environment is the scientific study of the chemical and biochemical phenomena that occur in natural places. A critical lesson learned in the past decade is that technology skillfully created to increase industrial production and to advance our standard of living is not an unqualified triumph over nature. We have come to realize that decisions to apply sophisticated technology must involve realistic evaluations of probable consequences in the vast interactive system that is our natural environment.

The environment must be regarded as an enormously complex system, one that includes not only the interdependent oceans, air and land, but also all forms of life and energy. The environmental problems that are our concern today-air and water pollution, waste disposal, conservation and renewal of natural resources, and lessening of the effects of natural disasters-cross geographic and political boundaries. They have legal and economic implications, they affect human lives; and they are characterized by enormous quantities of data.

Aditya Bhatnagar

Contents

Chapter 1

Forests in Environmental Protection

Forests play an important role in environmental protection. There is a long history of protection forests in mountain areas, where they help to prevent soil erosion, landslides and avalanches, and where they are important in maintaining the water quality of rivers draining forested catchments. Special silvicultural methods are required to ensure that these forests are maintained indefinitely. Forests also respond to environmental protection. A major issue is air pollution, which is known to have had significant impacts on some forests.

Air pollutants of concern include sulfur dioxide, hydrogen fluoride, heavy metals, and ozone. Control of these pollutants ultimately benefits forests. Forests have a major role to play in the protection of the global carbon cycle. They represent an important sink for atmospheric carbon dioxide, and conversion of forests to other land uses is one of the causes of the increase in atmospheric carbon dioxide concentrations. Reforestation and afforestation could contribute to reducing atmospheric carbon dioxide concentrations, and the use of biofuels could help to reduce demand for fossil fuels.

The relationship between forests and the environment has been recognized for more than a thousand years, yet forestry practices continue to cause damage to the environment in the form of soil erosion, water quality deterioration, and other adverse effects. Some of the earliest records of problems associated with the removal of forests come from Japan, where logging of the montane Cryptomeria japonica forests more than 1000 years ago was accompanied by increases in the incidence of flooding in low-lying areas.

Since then, forests have continued to be cut in headwater areas, often with disastrous consequences downstream. The value of forests as a means of environmental protection has, however, slowly been acknowledged. In areas where mass movements are common, and there is a relatively high population density, such as the European Alps and Japan, forests have been formally designated as having protection as their primary role. Elsewhere, the role of forest cover in protecting water sources and other values has also been recognized.

Links between forests and the atmosphere have been identified. As the control of pollutants and other substances altering the composition of the atmosphere becomes increasingly important, there has been a growing acknowledgement of the role that forests can play in global environmental protection. Forests are both affected by the pollutants and can themselves play a role in altering the atmospheric composition. Consequently, environmental protection measures taken to protect human health may have beneficial effects on forests.

Large-scale afforestation as a means of reducing atmospheric carbon dioxide concentrations has been discussed, and some countries (e.g., Australia) are already encouraging such a policy.

In this article, various aspects of the relationship between forests and environmental protection are examined. Examples are given from the boreal, temperate and tropical zones, in order to emphasize that environmental protection is an issue facing forestry in all parts of the world.

PROTECTION FORESTS

Protection forests have as their prime function the protection of the environment. They are common in mountainous areas—where they stabilize slopes, prevent avalanches, and protect water quality, and also in coastal areas, where they stabilize sand dunes. For example, in Austria, 741 452 ha of forestland have been classified as protection forest, representing 19.2% of the total forest area; these protection forests typically occur on steep (average gradient ca. 65%) slopes, and often occur at above 1800 m in altitude.

Elsewhere, the information on the amount of protection forest is very fragmented because the definition of protection forest varies between countries. For example, in Greece, 100% of the forests are classified as being managed primarily for soil protection, yet this does not preclude the annual removal of >2 million m^3 (overbark volume) of wood from the forest estate. In many tropical countries, certain forests have been formally designated as protection forests, but this designation is not always adhered to on the ground.

An example of the role of protection forests in maintaining water quality is provided by studies in the Oxapampa municipal watershed, near Cerro de Pasco, Peru. The area lies in an area characterized by tropical montane forest, and the landscape consists of steep slopes that are susceptible to erosion. Annual soil losses were compared between areas covered by natural forest, areas with a mixture of crops, pasture and forest, and areas of pasture only. The maximum suspended sediment concentrations reached 76, 226, and 771 mg/L, respectively, in the three land-use types during storm flows. Average annual soil losses from the three land-use types were estimated to be 121, 345 and 542 kg/ha, respectively.

Protection forests present particular problems for management. In Switzerland, in those protection forests where timber extraction is permitted, management has been described as "near-natural," involving either single-tree selection and continuous cover techniques or very small (<0.1 ha) patch cuts. In the French Alps, the ideal structure of protection forests dominated by Picea abies is believed to be uneven-aged stands with gaps. Management activities often aim to increase stand heterogeneity, and uneven-aged stand management has been adopted in some North American protection forests, such as the Quabbin Forest of central Massachusetts, which is an important source of Boston's water supply.

However, in some parts of the world, commercial felling within protection forests remains an important economic activity, and management is concentrated more on obtaining the best possible growth, while maintaining a forest cover. Thus, in the Russian Far East, forestry operations take place in the mixed, uneven-aged, protection forests.

The fellings are designed to reduce the proportion of Quercus mongolica, which is usually the dominant species in the overstory, and other broadleaved species. The management objective is to increase the proportion of Pinus koraiensis, a species that has a higher volume increment than Quercus mongolica. Such major interventions are not typical of European protection forests, but may be relatively common in areas where protection forests are still seen as an important source of timber.

Often, the frequency and extent of silvicultural intervention is strictly limited by law, a regulation that is based on the assumption that undisturbed forest ecosystems are stable. This is now known to be untrue: forest ecosystems are highly dynamic and subject to regular disturbance. As a disturbed protection forest may no longer fulfill its role, management is usually required to reduce the potential impacts of disturbances. Instances where protection forests lose their protective roles include those damaged by forest fires and by large-scale insect outbreaks, such as occurred in Switzerland as a result of major storms in 1990 and 1999. The 1990 storm caused major problems for foresters throughout Switzerland, and in other alpine countries. The amount of fresh dead wood was such that complete removal was impossible. This resulted in the build-up of bark beetle populations, further aggravating the instability of the forests. The problems may also have been exacerbated by the *laissez-faire* approach to management that has resulted in the stands containing a high proportion of very old trees and inadequate regeneration. In some cases, the associated high levels of dieback and mortality were attributed to air pollution, a convenient explanation that shifts the responsibility for forest instability away from the forest manager.

In another example (in the Swiss valley known as the Valais), the *Pinus sylvestris* forests situated on the slopes above the town of Visp have been severely impacted by *Tomicus* spp. This insect species normally does not directly cause the death of trees, as populations are generally insufficient to affect trees severely.

The cause of the severity of the pest outbreak in the Valais is uncertain, but may involve a combination of severe climatic stress, damage caused by fumes from a nearby industrial plant, very heavy mistletoe (*Viscum* album) infections, and the creation of good insect breeding habitat in the form of small groups of trees blown over by a severe windstorm. The effect is that the beetles have spread, and many trees have died. Salvage logging has been undertaken by conventional road-based logging, and also by the much more expensive helicopter logging.

The challenge now is to ensure that the gaps created by the salvage logging do not result in snow and debris avalanches that might endanger people living in the town below the forest. Assessments of the ecological stability of protection forests are still undertaken, and form an important step in the development of management prescriptions for such forests.

A recent case study of the Ban de Ville forest in Courmayeur, revealed many problems typical of alpine protection forests today. The 143 ha forest faces west, and is situated at altitudes between 1300 and 2300 m. It is dominated by *Picea abies*, although *Larix decidua* becomes increasingly common above 2100 m.

Two thirds of the forest was found to be "unstable," with identified problems including unsuitable stand composition, an oversimplified vertical structure and cover, the presence of high densities of wild ungulates, and the presence of forest pests (mainly the bark beetle *Ips typographus*) and diseases (mainly *Heterobasidion annosum*). Steps to reduce these problems included increasing the proportion of *Larix decidua* in the pure *Picea abies* stands and gradually establishing a multilayered, small group structure similar to that proposed for *Picea abies* protection forests in the French Alps. Reestablishment of protection forests following disturbance can present major problems. The forests grow close to the treeline, and are therefore, close to their ecological limit.

The microclimate in a forest is very different to that outside the forest, being more favourable to regeneration. Once the canopy cover is gone, the climatic conditions are much harsher, and it may be very difficult to establish a new tree cover. Consequently, it is almost impossible to reestablish the same species composition in a heavily disturbed protection forest. An example of such a system is the protection forests in the Urseren valley, near Andermatt, Switzerland. Here, the forests currently consist of *Picea abies* and *Larix decidua*, with some *Alnus viridis* scrub.

While regeneration is possible within the forest, regeneration would be impossible in the event that the forest cover was destroyed. A particular problem is sliding snow. In a mature forest, this process is limited by the presence of trees. When these are gone however, the buttressing effect is largely removed and snow avalanches can more easily occur.

Additionally, stems of young trees may be severed during avalanching. There are various strategies that can be adopted to reduce the problem. If the trees in a protection forest have died, they can be cut at a height of 50 cm to 1 m above the ground surface, leaving high stumps, and thus some buttressing potential. Alternatively, artificial barriers can be installed, although these may be costly. During the initial planting phase, a variety of techniques may be used to reduce snowslide, including stakes, tripods and horizontal logs 18–20 cm in diameter. In Europe, these barriers can be metal and concrete, but frequently, they are constructed from wood.

The wood from the surrounding forest can be used, but more often than not, such wood has a low mean life. For example, the heartwood of *Larix decidua* and *Pinus sylvestris* has a mean life of 8–15 years, which may be insufficient time for sufficient regeneration. The heartwood of *Castanea sativa* and *Robinia pseudoacacia* is more resistant to decay, but usually has to be brought in from elsewhere. It is also possible to ameliorate the impacts of sliding snow by planting trees that are less susceptible to this form of damage.

In extreme cases, very few trees are capable of growing, although *Pinus mugo* and *Alnus viridis* have been used successfully in the state of Bavaria in Germany. Clumps of *Larix decidua, Fagus sylvatica,* and Acer *pseudoplatanus* have also been used. *Sorbus* aria and *Sorbus* aucuparia are also fairly resistant to damage from snowsliding. These stabilize the snowpack, enabling the more sensitive Picea abies and Abies alba to be planted. Throughout British Columbia and southeast Alaska, red alder (Alnus rubra) vigorously colonizes snow avalanche tracks and is resistant to moderate-sized avalanches.

Another problem is the presence of very high populations of game (red deer, roe deer, chamois, and ibex) in some protection forests. These artificially high populations have arisen because of hunting policies that allow very high populations to exist (often involving feeding animals in winter and encouraging more palatable forage). In addition, recreation activities above the treeline are increasingly disturbing animals, so that they spend longer in the forest. The result is very high grazing pressure; in some areas this can severely affect forest regeneration.

This problem seems to occur throughout the alpine region, but is particularly apparent in Bavaria (Germany), Austria and Switzerland. It is actually a problem in many other types of European forest, and is not restricted to the Alps.

Protection forests do not only occur in mountainous areas. For example, in China, they are used extensively for protection against erosion in semiarid hill-gully systems. An example is provided by the "Three Norths" protection forest system in the northern part of the country. Plans for this system were passed by the State Council in 1978, with the first stage of development being completed in 1985. In establishing this system, account was taken of the ecological and economic benefits of the scheme, and the importance of rehabilitating and protecting the environment was given greater importance than timber production. In many areas covered by the scheme, forest cover has been increased dramatically.

For example, forest cover in one demonstration catchment on the loess plateau in Shanxi has been increased from 9.17% in 1977 to 24.4% in 1988. This is all the more remarkable given that the 8917 km^2 catchment has a population of 450 700 people. In another area, soil loss has been reduced by more than 95%, and runoff has been reduced by 91%. These changes have been accompanied by better agricultural planning, resulting in greater productivity of farmland, greater productivity of animal husbandry, and a 250% increase in per capita income.

Integrated development projects of this nature, where protection forests are one component of a series of improvements, are clearly of considerable potential. However, recent large-scale "soil and water conservation" projects on the loess plateau supported by World Bank funding are promoting the production of high cash value agricultural products (both crops and orchards) on large constructed terraces at the apparent expense of the long-term sustainability of semiarid soil resources. Another area where protection forests are of value is in coastal situations. Protection forests are used to stabilize sand dunes, but they are also used to protect muddy shorelines. In tropical and semitropical areas, mangroves are important as coastal protection forests. Very often, only a very small strip of forest remains. However, to fulfill protective functions, a mangrove forest needs to be quite wide. For example, in Indonesia, a greenbelt 614 m wide between the sea and forest fishponds has been suggested as necessary. Coastal protection forests have been used for a long time.

For example, the Etang-Sale forest was established on the island of Réunion in the 1870s in an attempt to stabilize the coastal dunes. The forest was planted with *Casuarina equisetifolia*, although more recently, a variety of other species has been used, including Acacia auriculiformis, Azadirachta indica, Eucalyptus camaldulensis and Khaya senegalensis, with Albizia lebbeck planted as an understory species to increase the vertical diversity. Coastal protection forests differ from mountain protection forests in that they are often planted on land that has never previously been forested. Consequently, exotic species are more often used than in mountain situations. For example, Pinus nigra (a pine species originating in southern Europe) has often been planted on sand dunes in Britain.

Many protection forests are designed to reduce the incidence of mass movements, particularly snow avalanches. However, forests may also help increase slope stability. This is because the roots of trees play an important part in increasing the strength of soils on slopes. If forests are removed, the roots will decay quite rapidly, with root strength normally at a minimum three to 15 years following clear cutting. The reduction in strength may be accompanied by landslides and other forms of slope instability.

In addition to landslides, the establishment of forests can also result in splash erosion. If there is very little vegetation cover under the forest, drip from the canopy may be sufficiently large and have sufficient energy to cause erosion of the soil. This is particularly true in some tropical plantations of *Tectona grandis*, especially where the leaf litter is removed. In such situations, it is possible to reduce the erosion by maintaining a ground cover, and by allowing the development of a thin organic horizon.

Despite these particular problems, there is a great deal of evidence that a well-managed forest will reduce the amount of surficial erosion, especially if it contains multiple strata (which help reduce the velocity of falling drips). The key to managing erosion in plantations is to maintain the infiltration capacity of the soils during the life cycle of the trees.

FORESTS BENEFIT FROM ENVIRONMENTAL PROTECTION

During the twentieth century, it become increasingly apparent that forests can be damaged by atmospheric pollution. A variety of pollutants may be involved, but sulfur dioxide, hydrogen fluoride, heavy metals, ozone, peroxyacetyl nitrate (PAN), and dust, can all be important. Early (late nineteenth and early twentieth century) records of pollution-induced damage were recorded in the immediate vicinities of industrial areas, such as the "Black Country" of England and the Ruhr of Germany.

At the same time, it was realized that the pollution was also harmful to humans, and as a result, controls on emissions were gradually introduced. These controls were not uniformly adopted. Although some controls were introduced in the nineteenth century, the majority date from the mid-twentieth century. Generally, urban areas in North America and western Europe were the first to receive attention, although there were notable exceptions. Eastern Europe, Russia, Asia, Africa, and South and Central America, mostly lagged behind in the introduction of controls. In many of these areas, priority continues to be given to jobs and development, with environmental protection taking second place. This has resulted in severe damage to forests in some industrial areas.

At the beginning of the twenty-first century, some of the worst damage is found in Russia and China. For example, widespread damage to forests and other vegetation has been reported from the Chongqing area of Sichuan Province in southwest China. The city of Chongqing is one of the biggest commercial and industrial centres of southwest China, and suffers as a result of the combustion of coal with a high-sulfur content. This combustion releases sulfur dioxide (SO_2), a toxic gas, into the atmosphere.

The concentrations of sulfur dioxide, and the resulting deposition of sulfate, result in some of the highest sulfur pollution levels in the world. Other pollutants in the area include mercury and fluoride. Many tree species, including *Pinus massoniana, Robinia pseudoacacia,* Eucalyptus *variegata,* Eucalyptus *robusta,* and *Quercus dentata,* have experienced a variety of forms of damage. Symptoms on the pines include tip necrosis of needles, yellow flecking, reduced foliar biomass, reduced needle length, premature needle abscission, branch dieback, reduced root growth, and reduced radial growth. Symptoms on the broadleaved trees include necrotic lesions, premature leaf abscission, and crown dieback. The damage is important to commercial plantations in the area, but also represents a threat to an area with particularly rich biodiversity, namely the Mt. Emei sacred mountain reserve.

Another area where forests have been severely damaged is the Kola Peninsula of northwest Russia. Here, there is a concentration of major smelters and other industrial plants, with particularly problematic locations being Nikel, Monchegorsk, Zapolyarny and Apatity. In the immediate vicinities of the smelters, the forests have been totally destroyed. There appears to be a sequence of ecosystem breakdown.

First, significant mortality occurs amongst the trees, probably as a result of toxic levels of sulfur dioxide. The increased volume of dead wood in the forest creates ideal conditions for fires, which kill any remaining trees. The sulfur dioxide and heavy metals may also kill most of the ground vegetation, exposing the soil to erosion. Eventually, there is little left other than bare rock. Reclamation in this arctic environment is extremely difficult and slow.

In South America, the Cubatao region in the Brazilian state of Sao Paulo is notorious for its pollution levels. The damage is of major concern as this area lies in the vegetation zone known as the Atlantic Rain Forest, which has a very high level of biological diversity but which has suffered major losses from forest conversion. As in other mountainous areas, the forest conversion and forest degradation has been accompanied by debris flows, mudslides and erosion, not only complicating reforestation, but also endangering the lives of people in the area. As in the Kola Peninsula, the damage appears to follow a clear progression.

First, the emergent trees are killed. Then the upper and middle layers of the forest canopy are affected, as are epiphytes growing in the canopy. Shrubs and herbs in the ground layer increase as the light levels reaching the ground increase. Not much is known about the sensitivity of individual species, but species in the Myrtaceae and Lauraceae families are amongst the first to disappear, whereas the Palmae (Arecaceae), Araceae, Melastomataceae, Moraceae, Ulmaceae, Urticaceae and Compositae (Asteraceae) seem to be more tolerant.

Such damage can also occur in western Europe and North America. A number of smelters in North America have caused damage to the surrounding forests. Examples of such point sources of pollution include Sudbury in Ontario, Trail and Kitimat in British Columbia, Spokane in Washington, and Copper Hills in Tennessee. In most of these cases, steps have been taken to reduce the pollution levels in the affected areas, and restoration efforts are underway. A particularly good example is provided by the recovery of forests in the Sudbury area. The techniques tried out at Sudbury could be applied to other parts of the world where restoration efforts are still in their infancy.

During the late 1970s and 1980s, there was considerable concern that the damage experienced around point sources might be occurring over much wider areas. This concern started in Germany, but rapidly spread to most European countries, and to North America. Surveys of forest condition were initiated, and these revealed that many trees were showing signs of what was considered to be poor health.

Pollution was blamed for the apparent problem, despite a lack of evidence to support such an assertion, and the term "forest decline" became widely used. Some scientists went so far as to predict that forests in Europe and eastern North America would have died by the year 2000. Huge amounts of research funds were directed to finding the cause of the problem, resulting in major progress in the understanding of tree physiology, but little progress in understanding the cause of the apparent poor health of the trees.

By the mid- to late-1990s, most research had stopped, and concern amongst environmentalists and others about pollution impacts on trees had been replaced by concerns about other issues. The reasons for this change in public perceptions are complex. Firstly, the extent of the problem appears to have been exaggerated. The index used to assess tree health was crown defoliation, measured in the field as crown density. This is a measure of the amount of light passing through a tree crown. It is assessed by a surveyor standing on the ground, looking up into the crown of the tree.

As such, it is a highly subjective measure, although various standards have been published that help increase the repeatability of the observations. When it was found that a large number of trees had defoliation values of between 25% and 60% (thereby receiving the classification "moderately damaged"), it was assumed that they were incapable of recovery, and likely to die. This has not been the case, and it seems that the crown density of trees can fluctuate markedly, with soil water availability playing a major role in determining the foliage area (as might be expected).

There are no records of crown condition prior to the late 1970s, so it is difficult to determine the long-term dynamics of this indicator. However, photographs taken at the turn of the century suggest that the crown density of trees was similar then to what it is today. Although the proportion of trees with some defoliation is reported to have increased throughout Europe over the past 20 years, there are numerous problems associated with the calibration of the time series, replacement of sample trees, and other technical difficulties, making interpretation of the series extremely difficult.

Research on forest decline produced conflicting results. While some claimed to have found the cause, others found no effect. In many cases, the symptoms were alleviated by fertilizing or liming. It seems likely that trees in any given area are subject to a number of different stresses. In some areas pollution may be important, in others it is not. Climatic stress, particularly drought, appears to have been important in triggering decline in many areas. Conversely, in eastern North America, climatic stress during winter appears to have played a major role in the onset of declines.

There is evidence for an interaction between environmental protection and protection forests. For example, in Switzerland, considerable concern has been expressed over the importance of vehicle emissions in destabilizing protection forests. Partly as a result of this concern, major investments have been made in building rail tunnels in areas where there is a combination of high traffic densities and sensitive protection forests. The 19 km tunnel between Klosters and Lavin in Switzerland is expected to carry 400 000 motor vehicles annually, and it is believed that this tunnel will help to preserve the Selfrauger protection forest. Whether this will actually be the case is debatable.

Some forms of pollution may be more important than previously recognized. For example, ozone (O_3) is a phytotoxic gas that may be affecting much larger areas of forest than have been identified so far. Some of the most important work on ozone and other constituents of photochemical smog has been done in the San Bernardino Mountains of southern California. This area lies inland of Los Angeles, and during the day, the emissions from vehicles in the Los Angeles area are swept inland by cool oceanic air masses being drawn into the interior to replace the rising hot desert air.

With time, these emissions evolve into a group of pollutants known as photochemical oxidants. These cause a number of direct and indirect effects, including chlorotic mottling of needles, branch dieback, tree mortality and ecosystem breakdown.

The evidence now is that the majority of forests in North America and western Europe are healthy and growing vigorously. Research on forest growth in Europe has revealed that forests are growing better today than they have during the rest of this century, a trend that many proponents of forest decline have had difficulty explaining. Despite these findings, there remains a hard core of individuals, particularly in forestry ministries, that remain convinced that air pollution is a major factor influencing the health of forests today.

THE ROLE OF FORESTS IN GLOBAL CYCLES

The Hydrological Cycle

An assertion that is often made is that forests increase rainfall. There is evidence that in a few cases, there may be small increases in rainfall associated with forest cover, but these increases are more than compensated for by increased evapotranspiration. The clearest evidence for increased rainfall comes from studies of the effects of climate change on the Amazon Basin, where total removal of the rainforest has been modeled to result in a decrease in rainfall of about 0.5 mm per day, particularly in the drier northeast.

For the whole Amazon Basin, this decrease would amount to a reduction in rainfall of about 6%. Outside tropical forests, there is very little evidence that forests increase the rate of rainfall, although they may increase the rate of fog drip.

The evidence relating forests to runoff is more convincing: forest cover in a catchment normally results in reduced runoff from that catchment, and removing that forest cover normally results in an increase in total runoff. Several processes are responsible, and the relationship varies with the annual precipitation, precipitation type (for example, whether or not winter snow occurs), and with forest type. Under wetter conditions, forests result in increased interception, and it is in these areas where the greatest effects of forest removal on total annual runoff occur.

However, in many wet temperate environments where major storms occur during the winter rainy season, the effect of this transpiration is minimal, and large peak flows do not increase appreciably following canopy removal. Under drier conditions, forests result in increased transpiration. Exceptions to this include cloud forests (where cloud water interception may exceed interception losses), forests in low-radiation and high rainfall environments (such as in coastal areas of New Zealand), areas with frequent fog (such as the coast of California), and very old forests (which may have lower rates of evapotranspiration than younger, vigorously growing forest).

Effects on the seasonal distribution of runoff vary from site to site, and are poorly documented. The effects of forests on the incidence of floods are also difficult to determine. The problems in Japan cited at the beginning of this article were clearly associated with deforestation in the headwaters of the catchments. However, a number of hydrological studies (in the US, UK, South Africa, and New Zealand) have shown no evidence of a link between storm flow and land use. In cases where there has been poor forest management—for example badly designed roads and ditches and excessive soil compaction—the size of floods may actually be increased. Most studies have shown that low, medium and high flows decrease with afforestation and increase with forest clearance, but the nature and degree of change of flows varies with site conditions.

The Carbon Cycle

With increasing concern about the rise in atmospheric carbon dioxide (CO_2) concentrations, it has been realized that forests play an important part in the global carbon cycle. Carbon released as a result of the conversion of forests to other forms of land use is an important cause of the increase in atmospheric CO_2, fossil fuels. In principle, afforestation and reforestation should be beneficial because young forests represent a net sink for atmospheric carbon. However, this effect is relatively short-lived and, as a forest reaches maturity, the rate of carbon uptake slows dramatically. This has led some to propose that older forests are replaced with younger forests, a proposal that has been vigorously objected to by the environmental movement. A critical point is what happens to any harvested timber.

If it is used in such a way that the carbon is released into the atmosphere quickly (e.g., it is burnt as fuelwood), then any carbon sequestration effect will only be temporary. However, if the carbon is not released for some time, then there is a potential for considerable carbon uptake. A more important process is the potential for wood to replace fossil fuels as a source of energy, although there are issues of pollution associated with wood-burning that need to be resolved. The potential benefits are being actively explored in several parts of the world.

At the biome level, there is considerable uncertainty about whether forests are currently acting as carbon sources or carbon sinks. For example, the boreal forest of Canada is believed to have been a net sink for carbon in the period 1900 to 1980, but since 1980, it has been a net source. A major factor in this change is the increased incidence of large-scale fires over the past 20 years. Global carbon budgets suffer from considerable uncertainty because of the lack of information about soil carbon budgets over much of the Earth's surface, and the lack of long-term data related to carbon storage. Some detailed work on carbon budget modeling has been conducted in the Brazilian Amazon.

The work of Houghton *et al.* (2000) has suggested that deforestation and abandonment of agricultural lands in the Brazilian Amazon was a source of 0.2 Pg C per year (1 Pg is 10^{15} g) over the period 1989–1998. Logging could be responsible for another 0.01 to 0.02 Pg C per year, and fires can increase the release of carbon by 0.4 Pg C in the years following a drought. Conversely, uptake by the natural ecosystems in the area approximately offsets the releases from land use change and fire, resulting in an approximate balance between sources and sinks.

Given the marginal economics of much forestry, there has been considerable interest in the idea that forestry could gain "carbon credits." The basic idea is that forest owners/managers would gain a financial benefit for sequestering carbon. The idea is present in the Kyoto Protocol, which lays out some mechanisms by which countries could reduce their net carbon emissions. Currently, this proposal is mired in controversy, as the negotiations surrounding the protocol have drawn attention to the lack of agreement on such terms as deforestation, afforestation and reforestation.

Even the meaning of the term forest is uncertain. There is concern that some countries will place emphasis on developing carbon sinks whereas many countries believe that the problem of global CO_2 concentrations must be addressed by reducing emissions. Not surprisingly, it is some of the countries with greatest potential for afforestation that support the sinks approach, whereas the European Union and other heavily industrialized areas support the emission reduction approach.

Forests in Restoration, Reclamation and Rehabilitation Projects

Many land areas have been severely degraded. Derelict urban areas, mine spoil, landfills, and abandoned industrial and agricultural land, are all now considered ready for redevelopment. In many cases, this redevelopment takes the form of conversion to "green" areas, such as parkland and even forest.

A distinction should be made between the concepts of restoration, reclamation and rehabilitation. Restoration involves the re-creation of the original forest conditions, including all species present in the original forest. As might be expected, it is extremely difficult.

Rehabilitation involves the use of some of the original forest species, but also allows the introduction of exotic species. In many cases, this enables a higher timber productivity to be achieved. Finally, reclamation refers simply to the reestablishment of a forest cover without any attempt to reconstruct the original species composition of the forest. Forests have a number of functions in redevelopment projects. They can act as barriers to noise and dust. They can provide a visually attractive feature in an otherwise barren landscape. They can also take up unwanted chemicals from the soil, although there is then a problem of disposing of the wood safely. Rehabilitation of degraded forests is also a major issue in many countries. A variety of factors can lead to degradation.

In Malaysia, the human-related causes of forest degradation include overcutting, conversion of forests to oil palm plantations and commercial agriculture, shifting cultivation, development projects such as dams, mining, and the removal of forest area from the permanent forest estate. Steps have been taken to restore, reclaim and rehabilitate forest lands, although the greatest efforts have gone into rehabilitation.

Rehabilitation efforts have encountered many problems, including the difficulty of getting the seeds of some tree species to germinate, inadequate knowledge of the ecology of many forest species, high mortality rates amongst seedlings, and great variability in the growth of seedlings raised from stumps and wildlings. Some of these problems could be overcome by greater investment in forest nurseries dealing with native species, and some success has been had with artificial regeneration using native species such as *Shorea parvifolia, Shorea leprosula, Neobalanocarpus hemii, Alstonia angustiloba, Dyera costulata* and *Azadirachta excelsa*.

While the concept of restoration is superficially attractive, it has a number of fundamental problems. Principal amongst these is to what state the forest will be restored. As more and more is learnt about forest dynamics, it has become clear that forests are not stable through time. Instead, they are highly dynamic. Consequently, it is important that any restoration efforts recognize this and create a forest where the processes of change are not inhibited in any way.

The traditional view of forestry is that it is oriented towards the production of timber. This is not the case for protection forests, where the principal product is environmental protection and security. Because of this, special management techniques are often required. For example, a priority for protection forests is the maintenance of forest cover. This means that the forest may be managed as a series of very small (<0.1 ha) patches, or by single-tree selection methods. In some cases, any form of management may be forbidden, but this can create problems if a natural disturbance occurs.

There are many assumptions and myths about the role of forests in environmental protection. Most of these have now been dispersed, but they occasionally reappear. Generally, it is inadvisable to make assumptions about the role of forests in environmental protection. Most case studies are precisely that, and their results should not be extrapolated to other forests or regions. Care should be taken not to overemphasize the role of forests in environmental protection. At the same time, greater recognition should be given to the important roles that forests play in protecting the environment.

Chapter 2

Environmental Protection Strategy of Arctic Ecosystems

There is a growing national and international appreciation of the importance of Arctic ecosystems and an increasing knowledge of global pollution and resulting environmental threats. The Arctic is highly sensitive to pollution and much of its human population and culture is directly dependent on the health of the region's ecosystems. Limited sunlight, ice cover that inhibits energy penetration, low mean and extreme temperatures, low species diversity and biological productivity and long-lived organisms with high lipid levels all contribute to the sensitivity of the Arctic ecosystem and cause it to be easily damaged. This vulnerability of the Arctic to pollution requires that action be taken now, or degradation may become irreversible. The governments of the Arctic countries have become increasingly aware of the need for, and their responsibility to combat these _ threats to the Arctic ecosystem.

On the initiative of Finland, the eight Arctic countries of USSR, USA, Sweden, Norway, Iceland, Finland, Denmark and Canada have met to prepare a strategy to protect the Arctic environment. The Arctic countries realise that the pollution problems of today do not respect national boundaries and that no state alone will be able to act effectively against environmental threats to the Arctic. They have also been moved by the international call for action expressed by the World Commission on Environment and Development as well as the concerns of the indigenous peoples living in the Arctic region. The Arctic countries with the participation of Arctic indigenous peoples have prepared this environmental protection Strategy. The strategy builds on the initiatives already taken nationally and by indigenous peoples to protect the Arctic environment. It is recognized that this Strategy, and its implementation, must incorporate the knowledge and culture of indigenous peoples. It is understood that the cultures and the continued existence of the indigenous peoples have been built on the sound stewardship of nature and its resources. The use of natural resources is an important activity of Arctic nations. Therefore, this Strategy should allow for sustainable economic development in the north so that such development does not have unacceptable ecological or cultural impacts.

The Strategy must also rely on the best scientific and technological advice that countries are able to produce and share. Arctic ecosystems are influenced and in some cases threatened by factors occurring also outside the Arctic. In turn, the Arctic also exerts an important influence on the global environment. The implementation of an Arctic Environmental Protection Strategy will therefore benefit both the Arctic countries and the world at large. The Strategy is also designed to guide development in a way that will safeguard the Arctic environment for future generations and in a manner that is compatible with nature. The Arctic countries are committed to international cooperation to ensure the protection of the Arctic environment and its sustainable and equitable development, while protecting the cultures of indigenous peoples. Only through careful stewardship by Arctic countries and Arctic peoples can environmental damage and degradation be prevented. These are the challenges which must be taken up in order to secure our common future.

The Strategy is comprised of a number of component parts, beginning with a statement of objectives. These objectives establish the broad direction in which the eight Arctic countries are intending to move. The objectives are accompanied by statements of principle which are designed to guide the actions of Arctic countries individually and collectively, as they move towards achievement of the objectives. The Strategy also describes the problems and priorities which the eight Arctic countries agree need to be addressed at this time. Tools, whether legal, scientific or administrative, are also reviewed in order to define appropriate mechanisms for implementation of the Strategy. This is particularly relevant to that section of the Strategy which defines the specific actions that the eight countries will undertake jointly or individually to deal with priority issues and pollution problems. The implementation of the Strategy will be carried out through national legislation and in accordance with international law, including customary international law as reflected in the 1982 United Nations Convention on the Law of the Sea. Finally, the Strategy outlines plans for future cooperation towards the implementation of the Strategy.

PROBLEMS AND PRIORITIES

At the first meeting in 1989 of the eight Arctic countries there was early recognition that many of the environmental problems that individual nations had been addressing, were in fact shared amongst the eight. To begin with, six specific pollution issues were identified as requiring attention. These issues were associated with persistent organic contaminants, oil, heavy metals, noise, radioactivity, and acidification. State of the Environment Reports were prepared on each of these topics and have been published separately. It was also agreed that these will be updated as necessary.

It was recognized that the ability to completely understand these issues was restricted by the lack of a comprehensive scientific data base and coordinated monitoring programme on the state of Arctic ecosystems. Furthermore, the potential impact of these specific pollutants on Arctic flora and fauna underlined the need to consider establishing a mechanism to facilitate a cooperative approach to their conservation. Other environmental problems including the depletion of the ozone layer and global warming were not addressed because they were already being considered in other fora. It was also determined that since the Arctic environment is particularly vulnerable to accidental discharges and uncontrolled releases of pollutants, enhanced mechanisms to address environmental emergencies in the Arctic were needed.

Persistent Organic Contaminants

The use and production of persistent organic contaminants (e.g. polychlorinated biphenyls (PCBs), DDT, hexachlorocyclohexane (HCH), chlordane and toxaphene has been stopped or restricted in some countries, however, many are still widely manufactured and used on a global basis. They are hazardous environmental contaminants due to their high stability and persistence in the environment, potential for bioaccumulation and high chronic toxicity, and the large quantities which have been released into the environment.

Although there are no major sources of these contaminants in the Arctic, they, nevertheless, reach the Arctic environment via long-range transport by rivers, the atmosphere and ocean currents from more industrialized centres, particularly Asia, Europe and North America. Due to the highly lipophilic nature of most chlorinated organic contaminants, they become concentrated in the fatty tissues of species in the Arctic food chain.

The highest levels of contaminants are therefore detected in the blubber and fat tissue of animals at the top of the food chain (e.g. polar bears, whales and seals). This is of particular concern in the Arctic because of the high level of consumption of lipid-rich wildlife foods by residents, resulting in a pathway of these contaminants to humans.

The presence of chlorinated organic contaminants has been reported in human populations throughout the world. The level of PCBs in breast milk samples collected from Inuit women in northern Quebec, was approximately five times higher than that of Caucasian women living in southern Quebec, Canada. The variable and generally sparse database on chlorinated organic contaminants in the Arctic prohibits for the most part, the determination of any spatial or temporal trends.

In Canadian studies, chlordane compound residue levels in polar bear fat have been reported to be four times higher in 1984 than in 1969, while levels of DDT did not change and other chlorinated contaminants measures were twice as high. Concentrations of chlorinated organic contaminants in the Arctic ecosystem are generally lower than in heavily polluted areas such as the Great Lakes, or the Baltic Sea. There are, however, some exceptions - the more volatile compounds (e.g. HCB, toxaphene) are often detected in the Arctic at concentrations similar to those In source regions.

Little is known about the potential effects of chlorinated organic contaminants on the ecosystem. However, there is evidence that a broad spectrum of contaminants is reaching the Arctic and there is sufficient toxicological data as well as field data to make reasonable extrapolations with regard to ecosystem consequences in the Arctic.

Chronic effects of chlorinated organic contamination observed in other regions (e.g. reproductive failure, bill and foot abnormalities, cancer) are of the most significance. The lower concentrations detected in the Arctic do not diminish the potential significance of their effects on ecosystem health.

Oil Pollution

The Arctic is one of the areas most vulnerable to adverse impacts from chronic and acute oil pollution. This is due to physical environmental conditions such as low temperature, periods with little or no light, ice cover etc. Low temperatures lead to reduced evaporation of the more volatile, toxic oil components.

Dark, cold winters in the Arctic lead to reduced ultraviolet radiation and biological decomposition of oil. In areas of drift ice, oil dispersal caused by wave action is also reduced. Oil in iced areas will be trapped between ice floes or under the ice, and only partly transported to the ice surface. These factors result in a generally slower decomposition of oil in the Arctic than in temperate regions. The period in which a particular oil spill can be harmful to wildlife is thus comparatively longer in the Arctic. The marginal ice zone is particularly vulnerable to oil pollution. A large part of the primary production in the Arctic, is found in this zone, which makes it extremely important for the whole Arctic ecosystem. Although there is no evidence that an oil spill reduces primary productivity to a significant degree, direct effects on marine life can be devastating, especially in the marginal ice zones. Feathers and fur contaminated by oil quickly lose their insulating properties, and the oil will often cause skin inflammation.

Both will lead to a negative energy balance of the affected animal. Ingested oil, in particular unweathered oil with a high content of volatile substances, can cause serious intoxication of birds and mammals. No studies indicate that any of these species tend to avoid oil spills. The amount of information available on oil spills in the Arctic, and probably the accuracy of the estimated quantities, varies considerably. Information on continuous discharges is scarce, and estimates of indirect oil transport (atmosphere, ocean currents, and rivers) have not been available.

Order of magnitude calculations show that river transport is the main contribution of oil pollution to the Arctic (estimated at 200,000 metric tons per annum). The highest risk of oil spills is connected with transportation activities and production of oil as well as to a lesser degree, exploration activities. Their occurrence will depend on the level of activity in the Arctic, the technical standards of the activity and the preventative measures taken.

The physical constraints caused by Arctic conditions imply particular technological challenges regarding oil spill clean up. Effective methods and techniques for containing and cleaning up oil spills from water and ice are currently limited. The available information on ambient oil pollution in the Arctic is scarce. More information is needed, obtained with standardized methodologies in order to have comparable data for the whole region, with special emphasis on fluvial inputs and concentrations in surface marine waters.

Heavy Metals

Levels of heavy metals have been found in the air, precipitation, ocean waters, soils, rivers, lakes and bottom sediments of the Arctic as well as in marine, freshwater and terrestrial biota. These levels occur as a result of natural phenomena as well as from regional sources and global transport. Heavy metal concentrations in air and precipitation are mainly due to long-range atmospheric transport from industrial centres resulting in a deposition of heavy metals on vegetation, snow, and the sea which generally decreases from south to north. To a lesser extent discharges from local mining operations and the methylation of inorganic mercury often associated with large scale impoundments of water in previously vegetated areas (i.e. hydroelectric developments), also account for elevated heavy metal concentrations. Canadian and Finnish studies indicate methyl mercury levels in fish rise measurably after the flooding of new reservoirs, depending upon the amount of organic material present.

The temporal trends of long-range heavy metal pollution of the Arctic particularly mercury, cadmium, lead, arsenic and nickel have been determined by analyzing ice cores from glaciers. There has been an increasing trend since the middle of the 19th century and a sharp increase in the 20th century. Recent analyses of vegetation seem to indicate that a decrease may now be occurring.

The concentrations of heavy metals in lakes and rivers are generally higher than in Arctic sea water. A decrease in pH caused by acid precipitation increases the dissolution rates of heavy metals which may increase the rate of accumulation in the biota. In the Arctic marine environment the concentrations of heavy metals in water are low compared to more southerly latitudes. However, the concentrations in biota increase in the food chain, and in the top level predators such as seals and whales the concentrations, especially of cadmium, increase to levels much higher than observed in other areas.

For example, in some Canadian studies, cadmium levels in narwhal kidney were among the highest ever reported in marine mammals. This build-up is probably due to naturally occurring phenomena, but such occurrences make increases in the concentrations of heavy metals in the Arctic marine environment as a result of industrial sources more problematic than elsewhere. The high concentrations of heavy metals in marine mammals and some bird species constitute a problem in districts where tissues from such animals constitute a significant part of the diet. Thus increased concentrations of mercury have been found in Greenlanders from hunting districts. Elevated levels of mercury have also been found in the Native populations of Northern Quebec, Canada.

Noise

The waters of the Arctic region are a unique noise environment mainly due to the presence of ice. The ambient noise is strongly influenced by the dynamic processes of ice formation, melt, deformation and movement. This situation is different from ice free waters. In periods where ice cracking and wind noise are absent, areas covered by shore-fast ice are among the quietest underwater environments. Human activities create noise types and levels, which may disturb marine mammals, or mask the "natural" sounds of importance to those mammals. Some types of noise may affect fish as well as marine mammals. There are a number of serious gaps in our knowledge of the effects of underwater noise on marine mammals, including the inability to assess the effect of repeated noise exposure on stocks. There is considerable evidence that most types of disturbance do not cause mortality.

However, some noisy activities, including low level overflights by aircraft, near seals and walrus at haul out sites can cause mortality through stampedes or abandonment. Many marine mammals seem able to adapt to or at least tolerate many types of disturbances or increased noise levels. However the scarcity of direct evidence of serious consequences from disturbances does not necessarily mean that marine mammals are not stressed or affected in some other way.

Noise from human activities may cause short-term or long-term behavioural reactions and temporary displacement of various marine mammals. The biological significance of most of these reactions is unknown. Moving sound sources, notably boats and aircraft, seem to be more disturbing than stationary sources, e.g. dredges and drillships. The effects on fish and wildlife of cumulative exposure to noise are largely unknown.

Radioactivity

There have been two major causes of radioactive contamination affecting the Arctic region: atmospheric nuclear-weapons testing during the 1950s and 1960s and the accident at the Chernobyl nuclear power plant in 1986. Of greatest concern are the long-lived radionuclides, including Strontium-90 (29 year half-life) and Cesium-137 (30 year half-life). Studies have shown that these fallout derived radionuclides are efficiently retained by surface vegetation, especially lichen, in this nutrient-poor environment and are biologically recycled in Arctic ecosystems.

As a result, those indigenous peoples and local populations consuming as their main food caribou or reindeer meat with elevated levels of radio-cesium, may have accumulated higher levels. Other radioactive threats to the environment exist, e.g. accidental discharges which are of a biological significance associated with nuclear power sources and transport, storage and disposal of radioactive waste. When considering the total radiation dose, attention should be given to the radiation from man-made sources and to natural radiation. A number of bilateral and multilateral arrangements, including those with the International Atomic Energy Agency (IAEA), address issues related to exchange of information, early notification of radiation release, emergency preparedness and response to nuclear accidents and transboundary movement of radioactive materials.

Acidification

The most important acidifying substances are sulphur and nitrogen compounds emitted mainly by vehicles, industrial activities and coal and oil based power plants. Long-range transport is the most important factor influencing the air quality in the Arctic, especially in winter. The sulphur and nitrogen emission from industrial activities in the Arctic is also a considerable factor. Until now, little emphasis has been placed on the effects of acid deposition on Arctic ecosystems.

Furthermore, knowledge derived from studies in temperate zones is not directly relevant to the Arctic. One of the most well known examples of a problem associated with _ acidity in the Arctic is the Arctic haze phenomenon produced from acid pollutant aerosols. Arctic haze has been under intense study and much is known about its nature, distribution and composition.

Acidification is evolving into a prominent environmental problem around certain northern industrial centres. In northern FennoScandia, in the northwestern parts of the Soviet Union and in the eastern parts of Canada, natural factors increase the sensitivity to acidification and anthropogenic impacts have extended through the whole area.

The interaction between acidic deposition and the soils of different ecosystems is an important component of the acidification process. A continuous excessive acid load leads to the mobilization of aluminum and heavy metals.

The combined effects of acid deposition and the stresses already induced by the harsh climate increase the possibility of vegetation damage in the Arctic. Critical loads, rates of acidification, and conditions influencing cold climate environments need more detailed regional monitoring and research. In general, northern ecosystems are under greater stress than temperate ecosystems.

INTERNATIONAL MECHANISMS FOR THE PROTECTION OF THE ARCTIC ENVIRONMENT

Before determining the specific actions required to protect the Arctic environment, a review of existing and proposed international and bilateral agreements and policy declarations pertaining to the Arctic environment has been completed. This review has revealed the existence of a number of mechanisms that may be employed to protect the Arctic environment, and provides a useful tool for the implementation of the Strategy. The List of Major International Instruments and Policy Declarations Pertaining to the Arctic Environment has been distributed and will be periodically updated. Through the more specific studies of the six priority areas, some important gaps have been identified.

Persistent Organic Contaminants

International legal instruments which currently control air pollution are not specifically directed at limiting the emission of persistent organic contaminants and related contaminants. Work is underway within the UN ECE Convention on Long-Range Transboufldary Air Pollution (LRTAP) to review the problem and identify specific control actions.

Any new international legal instrument would need to target contaminant reductions in the industrialized areas of Asia, Europe and North America. Consequently, proper management and protection of the Arctic ecosystem from the effects of these contaminants will require cooperation through effective bilateral and multilateral agreements among both circumpolar and non-circumpolar nations.

Oil Pollution

The main international instruments relevant inter alia to oil pollution in the Arctic are:

- The 1969 international Convention on ivil Liability for Oil Pollution Damage (Convention);
- The 1969 InternatiOnal Convention relating to Intervention on the High Seas in Cases of Oil Pollution Casualties (Convention on Intervention);
- The 1971 international Convention on the Establishment of an International Fund for Compensation for Oil Pollution Damage (Fund Convention);
- The 1972 Convention on the Prevention of Marine Pollution by Dumping of Wastes and Other Matter (London Dumping Convention);
- The 1974 Convention on the Prevention of Marine Pollution —from Land-Based Sources (Paris Convention);
- The International Convention for the Prevention Of Pollution from Ships;
- The 1982 United Nations Convention on the Law of the Sea (UNCLOS)(not yet in force); viii) the International Convention on Oil Pollution Preparedness, Response and Cooperation, 1990 (not yet in force).

Certain issues with respect to oil pollution in the Arctic are insufficiently covered by the existing international agreements and conventions. For example, some instruments are limited in their application and only partially apply to the Arctic region. There is a need to consider the possibility of extending the geographic scope of these instruments.

The provisions of the various instruments also need to be further assessed to determine their adequacy under Arctic conditions taking into account the particularly vulnerable nature of the region. Strict standards in the transportation of oil in the Arctic are needed. Such standards should be developed under the framework of the International Maritime Organization (IMO). Arctic countries should consider becoming parties to or applying the relevant principles of the various existing conventions and agreements on oil pollution management.

Heavy Metals

There are few multilateral conventions and bilateral agreements which deal with heavy metal inputs in the environment. The UN ECE LRTAP Convention is one of the major international conventions which limits harmful atmospheric emissions.

Work is currently underway within this forum to identify and decide upon specific control actions to deal with heavy metals. Control of discharges of heavy metals to the marine environment are governed by both the 1972 Convention for the Prevention of Marine Pollution by Dumping from Ships and Aircraft (Oslo Convention) and the 1974 Convention on the Prevention of Marine Pollution from Land-based Sources.

The UNCLOS provides opportunities for controlling the discharge of harmful substances e.g. heavy metals. There are only a few bilateral arrangements between the Arctic countries controlling these substances.

Noise

Existing legal instruments do not address the effects of noise on the Arctic ecosystem. There may be a need for Arctic countries to agree on the adoption of procedures to ensure that in the planning and conduct of activities in the Arctic, measures are taken to facilitate the adequate monitoring of the potential disturbance from noise including the verification of predicted effects and the identification of any unforseen effects. Such evaluations should ensure that environmental protection measures are given due consideration.

Radioactivity

The 1986 international Atomic Energy Agency (IAEA) Convention on Early Notification of a Nuclear Accident, provides an adequate mechanism for cooperation and exchange of information applicable in the Arctic region. In addition, there are a number of bilateral and multilateral agreements in existence which provide for such events as the early notification of accidental radioactive discharges (e.g. between Finland and other Nordic countries, Finland and the USSR, 1987).

With respect to emergency preparedness and assistance, the IAEA Convention on Assistance in the Case of a Nuclear Accident or Radiological Emergency was established to facilitate prompt international assistance when requested in the event of a nuclear accident. Additionally, a number of bilateral and multilateral agreements have been concluded between the Arctic countries that both supplement and provide greater precision to the IAEA arrangements. These should be examined to ensure that the specific environmental conditions of the Arctic are addressed. Furthermore, consideration should be given to practical mechanisms between national authorities to implement the coordination of emergency measures within the existing international legal framework of the IAEA.

Acidification

The Arctic is exposed to the long range transport of acidifying — substances from various sources. A number of control measures have been introduced at the United Nations Economic Commission for Europe (ECE) and bilaterally. Although, in general, these measures do not contain provisions relating specifically to the Arctic region, they do refer to emission sources that affect the Arctic.

Radioactivity The 1986 international Atomic Energy Agency (IAEA) Convention on Early Notification of a Nuclear Accident, provides an adequate mechanism for cooperation and exchange of information applicable in the Arctic region. In addition, there are a number of bilateral and multilateral agreements in existence which provide for such events as the early notification of accidental radioactive discharges (e.g. between Finland and other Nordic countries, Finland and the USSR, 1987).

With respect to emergency preparedness and assistance, the IAEA Convention on Assistance in the Case of a Nuclear Accident or Radiological Emergency was established to facilitate prompt international assistance when requested in the event of a nuclear accident. Additionally, a number of bilateral and multilateral agreements have been concluded between the Arctic countries that both supplement and provide greater precision to the IAEA arrangements. These should be examined to ensure that the specific environmental conditions of the Arctic are addressed. Furthermore, consideration should be given to practical mechanisms between national authorities to implement the coordination of emergency measures within the existing international legal framework of the IAEA.

Acidification

The Arctic is exposed to the long range transport of acidifying — substances from various sources. A number of control measures have been introduced at the United Nations Economic Commission for Europe (ECE) and bilaterally. Although, in general, these measures do not contain provisions relating specifically to the Arctic region, they do refer to emission sources that affect the Arctic.

ACTIONS

The eight Arctic countries agree to proceed cooperatively with the following action plan. These commitments will begin the process of addressing the serious environmental issues identified and assessed through the preparation of specific state of environment reports. These issues will require regular updates for evaluation by the eight Arctic countries on the progress being made, and to advise on possible new courses of action.

Persistent Organic Contaminants

- In order to further define the likely sources, pathways, sinks and effects of these pollutants, and to expand the data base to cover all the component parts of the Arctic environment, the eight Arctic countries will undertake cooperative monitoring (AMAP) and research related to the problem of persistent organic contaminants in the Arctic ecosystem.
- The Arctic countries will consider the feasibility of developing national inventories on the production, use, and emissions of persistent organic contaminants (e.g. pesticides) to be collected, and made available and summarized in the state of the Arctic environment reports.
- The Arctic countries will also address the problem of persistent organic contaminants under existing or proposed international agreements and will review other mechanisms to advance this issue in other international fora.
- In order to achieve an early reduction in the movement of

persistent organic contaminants into the Arctic environment, the eight Arctic countries will support the process now under way within the UN ECE LRTAP Convention to further define the problem and to develope proposals for international action on the control of these substances under the Convention. Those Arctic countries which are partners to the Paris and Helsinki Conventions will actively support ongoing inventory and assessment work under those conventions.

- The Arctic countries agree to implement measures to reduce and/or control the use of the following polluting substances: chlordane, DDT, toxaphene and PCBs. Those Arctic countries which have not already done so, also recognize that the elimination of the problem of persistent organic contaminants in the Arctic may also require controls on the production of these substances.
- The Arctic countries will review the situation with regard to other persistent organic contaminants with a view to establishing priorities and timetables for a programme of emission elimination or control in cooperation with other international fora.

Oil Pollution

- In order to achieve better documentation of the level of oil pollution in the Arctic environment, the initiation of monitoring of hydrocarbons as a part of the AMAP, will play an important role.
- There is also a need to consider establishment of a reporting system on discharges and spills, with regard to provide adequate documentation on the pollution threat, in the Arctic.
- The elements agreed upon in Section 8, Protection of the Arctic Marine Environment and Section 9, Emergency Prevention. Preparedness and Response will comprise the basis for further cooperation in preventing and combatting oil pollution.
- The Arctic countries agree to take measures as soon as possible to adhere to the strictest relevant international standards within the conventions, to which the countries are parties, regarding discharges irrespective of origin.
- The Arctic countries agree to undertake joint actions in relevant international fora to further strengthen recognition of the particularly sensitive character of icecovered parts of the Arctic Ocean.

Heavy Metals

- An improved understanding of the dynamics of heavy metals

in the Arctic ecosystem is required. The countries will undertake a programme of coordinated monitoring (AMAP) and research to identify sources, pathways and sinks of heavy metals; spatial and temporal trends; and, ecological effects with special emphasis on human health effects.

- The eight Arctic countries agree to implement measures to control conditions that lead to the release of heavy metals by industrial activities including as appropriate the implementation of best available technology and other concerted actions in accordance with appropriate international agreements (e.g. UN ECE LRTAP Convention).

Noise

- The effects of noise associated with Arctic marine and terrestrial projects should be evaluated as part of the project planning and approval processes, and if significant adverse noise effects on the specific components of Arctic ecosystems are predicted, then measures should be implemented to avoid or mitigate the impact.
- Efforts should be made to improve the knowledge on marine mammal auditory function, communication and behaviour and the current noise exposure assessment techniques. For specific project evaluations, site-specific data should be addressed before and during the evaluation. This includes determining how much exposure migrating stocks are encountering throughout the year.

Radioactivity

- AMAP should address radioactivity. Common standards and techniques for monitoring and analysis, consistent with IAEA standards and technology should be developed.
- Future monitoring and health assessments should consider the effects from exposure to radiation from man-made sources together with natural or background radiation.
- Further consideration should be given to the development of more specific measures, consistent within the international legal framework of IAE procedures, for cooperation amongst Arctic countries to deal with emergencies caused by the accidental release of radioactive substances and to provide mutual assistance in the harsh Arctic environment.
- All relevant data concerning previous studies and measurements should be collated in the existing relevant data bases of which information should be exchanged between the governments and institutions concerned.

Acidification

- Regional Arctic research programs should be developed to assess the current loadings and potential effects of acid deposition on representative sensitive Arctic ecosystems. Special attention should be given to those regions or ecosystems for which existing data or assessments suggest that there is or is likely to be an acidification problem.
- Consideration should be given to expanding deposition monitoring programs, within the framework of AMAP and existing networks such as the ECE/EMEP deposition monitoring network, to encompass measurement of acid deposition in the Arctic. Emphasis should also be placed on measuring dry deposition.
- Emphasis should be placed on defining critical loads and setting target loads for sensitive Arctic ecosystems. In the event that these target loads are being exceeded, steps should be taken to reduce those emissions contributing to the problem, in accordance with international agreements such as the ECE LRTAP Convention. Reduction of emissions of sulphur and nitrogen should be sought by, *inter alia.* implementing the use of the best available technology.

ARCTIC MONITORING AND ASSESSMENT PROGRAMME

The eight Arctic countries recognize that the Arctic region represents one of the relatively pristine areas on earth. It is therefore of great importance to preserve and to protect the Arctic. Measurements in the Arctic indicate that pollutants originating from anthropogenic activities in the mid-latitudes are transported to the Arctic by atmospheric processes, ocean currents and rivers, and that pollutants are deposited and accumulated in the Arctic environment and its ecosystems.

Exploitation of natural resources, and concomitant urban and industrial expansion within the Arctic region, also contribute to the degradation of the Arctic environment and affect the living conditions for the people of the region. Distinguishing human-induced changes from changes caused by natural phenomena in the Arctic will require monitoring of selected key indicators of the Arctic Environment.

Therefore, the eight Arctic countries have agreed to promote development of an Arctic Monitoring and Assessment Programme (AMAP) in order to understand and document these changes and so that the monitoring results may be used to anticipate adverse biological, chemical and physical changes to the ecosystem and to prevent, minimize and mitigate these adverse effects. The primary objective of the AMAP is the measurement of the levels of anthropogenic pollutants and the assessment of their effects in relevant component parts of the Arctic environment. The assessments should be presented in status reports to relevant fora as a basis for necessary steps to be taken to reduce the pollution. Two of the most significant threats to the present Arctic environment may come from climate change, induced by global warming, and the effects of stratospheric ozone depletion.

Programs to detect and determine the causes and effects of climate change and ozone depletion are to a large extent being developed by other international groupings and in other fora. It is important for AMAP to be aware of these programs and to develope links with them from an Arctic perspective in order to encourage and facilitate an Arctic component in climate programs. Data obtained for assessing climate change will provide important inputs to the AMAP dataset. In turn, AMAP data will be relevant to -climate change programs in the Arctic.

The pollution data available from the Arctic region are with a few exceptions based on research programs performed within limited subject areas by national programs and not supported by bilateral or international cooperation. There is an urgent need for cooperation among local and regional efforts and global programs in order to obtain better documentation on the environmental situation in the Arctic especially with regard to longrange air and marine pollution. From the outset, the AMAP should as far as possible be based on existing programs. The programme should be initiated in a step by step fashion as indicated in the proposal for the AMAP.

Actions

- Distinguishing human-induced changes from changes caused by natural phenomena in the Arctic will require estimates and regular reporting by the Arctic countries of contaminant emissions and discharges, including accidental discharges, as well as transport and deposition. In addition monitoring of deposition and selected key indicators of the Arctic biological environment, are required. The eight Arctic countries should therefore agree to establish an Arctic Monitoring and Assessment Programme (AMAP) to fulfill these monitoring objectives.
- The AMAP should be implemented through the establishment of an Arctic Monitoring and Assessment Task Force and a small secretariat, established by the Government of Norway.
- AMAP should as far as possible build upon existing programs.

Thus, one of the important tasks of the AMAP will be to review and coordinate existing national programs, establish a data directory, and to develope these programs when appropriate in an international framework.

- As an initial priority, the AMAP should focus on persistent organic contaminants and on selected heavy metals and radionuclides, and ultimately to monitor ecological indicators to provide a basis for assessments of the status of Arctic ecosystems.
- The eight Arctic countries will receive regular State of the Arctic Environment Reports summarizing the results of the AMAP.

As a result of these actions, the Arctic Monitoring and Assessment Programme will provide information for:

- Integrated assessment reports on status and trends in the condition of Arctic ecosystem;
- Identifying possible causes for changing conditions;
- Detecting emerging problems, their possible causes, and _ the potential risk to Arctic ecosystems including indigenous peoples and other Arctic residents;
- Recommending actions required to reduce risks to Arctic ecosystem

PROTECTION OF THE ARCTIC MARINE ENVIRONMENT

The eight Arctic Countries recognize their particular interests and responsibilities as neighbouring countries in the Arctic, and emphasize the need to take preventive measures directly or through competent international organizations, consistent in particular with the 1982 United Nations Convention on the Law of the Sea regarding marine pollution in the Arctic, irrespective of origin. To this end the Arctic countries agree to:

- Apply the principles concerning the protection and preservation of the Marine Environment as reflected in the 1982 United Nations Convention on the Law of the Sea, and, in accordance with the continuing development of international environmental law, to further strengthen rules in order to protect the Arctic;
- Take measures as soon as possible to adhere to the strictest relevant international standards within the conventions, to which the countries are parties, regarding discharges irrespective of origin;
- Undertake joint actions in relevant international fora to further strengthen recognition of the particularly sensitive character of ice-covered parts of the Arctic Ocean;

- Review, in accordance with the general aims of this environmental Strategy, the relevance to the Arctic of international instruments connected with the protection of the marine environment, with the aim that all Arctic countries accede, where appropriate, to the instruments, or apply the principles and regulations embodied therein;
- Jointly support the appropriate initiatives of international organizations in developing mandatory standards in order to improve the protection against accidental pollution affecting the marine environment, and actively ensure application of such standards;
- Carry out studies of pollution in the monitoring activities within AMAP.

EMERGENCY PREVENTION, PREPAREDNESS AND RESPONSE

At the same time as the Arctic is exhibiting signs of serious contamination from pollutants carried via long range transport from mid latitudes, there has been an increase in development activities and shipping within the Arctic. These activities can have serious environmental consequences in the Arctic as a result of accidents leading, *inter alia*, to spills and discharges of oil and other harmful substances. The vulnerability of the Arctic ecosystem to these sudden intrusions will be variable.

Some limited mapping of areas sensitive to oil spills has been conducted but more remains to be done. The relative hazard/risk associated with different activities is also not well documented, nor is the geographic distribution of high risk activities. There are a number of bilateral, regional and global arrangements which presentl; exist to deal with accidental pollution, such as the 1983 Canada-Denmark Agreement for Cooperating relating to the Marine Environment, the 1971 Agreement between Denmark, Finland, Norway and Sweden on Cooperation on Oil Pollution and the 1990 International Convention on Oil Pollution Preparedness, Response and Cooperation.

There are other multilateral conventions related to nuclear accidents or radiological emergencies supplemented by bilateral agreements on the exchange of information and reporting relative to nuclear plants and events. The UN ECE has started work on an international convention, on the prevention and control of the transboundary effects of industrial accidents. A part of the work is the establishment or reinforcement of regional and subregional mechanisms for response, assistance and exchange of information on environmental emergencies.

Actions

The Arctic countries agree to the following framework for taking early cooperative action on emergency prevention, preparedness and response in the Arctic. They will take steps to review existing bilateral and multilateral arrangements in order to evaluate the adequacy of the geographical coverage of the Arctic regions by cooperative agreements. They will also take steps to convene a meeting of experts to consider and recommend the necessary system of cooperation, which could include, *inter alia*, the following elements:

- Actions to respond to significant accidental pollution from any source;
- Coordination and harmonization of preventive policies, strategies and measures;
- Establishment of a system for early notification in the event of significant accidental pollution or an imminent threat of such pollution;
- Assessment of the risks for significant accidental pollution and of the adverse effects in such cases so as to enable the parties to take the necessary preventive, preparedness and response measures;
- Inclusion of studies on effects of accidental pollution in conjunction with the monitoring activities of AMAP;
- Cooperation in the conduct of research into and development of methods and technologies for prevention of, preparedness for and response to significant accidental pollution in the Arctic;
- Cooperation in developing a system for exchange of information on research and new developments regarding methods and technologies on response in the Arctic;
- Exchange of information on legislative and administrative measures as well as policies;
- Measures for providing information to the public and public participation;
- Further enhance regional bilateral and multilateral cooperation in the Arctic regarding prevention, preparedness and response by developing, as appropriate, contingency plans, training programs, as well as other measures to facilitate assistance to the parties, in particular mutual assistance for efficient emergency response in the event of significant accidental pollution, or the imminent threat of such pollution.

CONSERVATION OF ARCTIC FLORA AND FAUNA

The health of Arctic flora and fauna is a key concern of the Arctic countries. These flora and fauna assume special significance in this region since they are an essential factor helping to define the culture and survival of the people living there. Although isolated geographically from the industrialized temperate regions of the globe, it has now been amply demonstrated that this has not excluded Arctic flora and fauna from the negative consequences of human activities in mid latitudes.

The impacts on the Arctic have escalated over the past several decades and both scientific and traditional knowledge have been pointing to the danger signals. Many of these concerns are enumerated in the six Arctic State of the Environment reports. They confirm that Arctic flora, fauna and their habitats are being threatened by large scale economic development projects; long range transport of pollutants; and degradation of habitats.

The problems facing Arctic flora, fauna and habitats are not confined to any one country but are circumpolar in nature. Furthermore, because of the uniqueness of Arctic ecosystems, strategies to deal with these problems will differ from those of other regions. Several multilateral and bilateral agreements which pertain to the conservation of Arctic flora and fauna and their habitats are currently in existence. Most however, have been designed to be universally applicable to, or to apply to, a wider geographical area than the Arctic.

Only the Agreement on Conservation of Polar Bears and some individual provisions in other agreements provide a specific Arctic focus. The eight Arctic countries should therefore seek to create a distinct forum for scientists, indigenous peoples and conservation managers engaged in Arctic flora, fauna and habitat related activities to exchange data and information on issues such as shared species and habitats and to collaborate, as appropriate, for more effective research, sustainable utilization and conservation.

Actions

The eight Arctic countries are mindful of the need to conserve Arctic flora and fauna and their habitats in their natural diversity, and protect these resources from the pollution threats I described in this Arctic Environmental Protection Strategy. They recognize the special relationship and importance of Arctic flora and fauna and their habitats to indigenous peoples.

The countries also recognize the benefits to be gained from sharing scientific and management information, traditional knowledge, and other data with respect to Arctic flora and fauna and their habitats. With due regard to existing international cooperation, and in an effort to improve research and information aimed at protecting these resources and their habitats from pollution and environmental degradation, they have reached the following understanding:

- The eight Arctic countries will cooperate for the conservation of Arctic flora and fauna, their diversity, and their habitats. Such cooperation shall include, *inter alia*, exchanges of research and management information and data, and coordination of research, on the following:
  - Arctic species, their health and habitats;
  - The laws, regulations and practices of the parties with respect to the conservation and management of such species;
  - The importance and relationship to, and use of, such species by indigenous peoples and the unique contribution of indigenous peoples to the stewardship of nature and its resources;
- Each country will provide to the other countries, as appropriate, such information, publications, and/or documents as may be agreed under the terms of the Strategy;
- The eight Arctic countries will seek to develope other forms of cooperation, including exchanges of experts, of traditional knowledge, and of other data, as well as engaging in joint projects, bilateral or multilateral meetings, symposia and joint publications, to meet the intent of this Strategy;
- The eight Arctic countries will each seek to develope more effective laws, regulations and practices for the conservation of Arctic flora and fauna, their diversity, and their habitats in close cooperation with Arctic indigenous peoples;
- The eight Arctic countries agree to establish a mechanism for furthering the following aims in close cooperation with Arctic indigenous peoples:
  - Promoting and facilitating exchanges of information and personnel as provided for in this Strategy;
  - Making recommendations with respect to the priorities, the orientation and the nature of research and monitoring programs of the Arctic Countries;
  - Proposing strategies for enhanced conservation of Arctic species and their habitats; and
  - Regularly compiling and disseminating information on activities regarding the conservation of Arctic flora and fauna.
- The eight Arctic countries will consult, as deemed appropriate with the International Arctic Science Committee and other

bodies on any matter that falls within the scope of this Strategy;

- By October 1991 each Country will identify to the others its national agency designated to Coordinate the cooperation envisaged by this section;
- The Countries agree that the terms and conditions of the cooperation and exchanges provided for in this section will be subject to the laws and regulations of the Countries;
- Each country will make its best efforts to provide resources adequate to carry out its responsibilities under this section. It is understood that the ability of each country to carry out activities is subject to the availability of funds, and that countries will seek to ensure long-term funding for necessary projects.

FURTHER COOPERATION

Continuity and further cooperation are essential for increasing the protection of the Arctic environment. In order to ensure, this continuity and cooperation, the eight Arctic countries agree to hold regular Meetings on the Arctic Environment. The date and venue of the next meeting will be agreed upon at the preceding meeting. Decisions on the agenda and participation of observers will be made and communicated to interested parties in advance of the meeting.

The decision to invite observers should be based on a pragmatic and functional evaluation of their involvement in and contribution to Arctic environmental questions. In order to facilitate the participation of Arctic indigenous peoples the following organizations will be invited as observers: the Inuit Circumpolar Conference, the Nordic Saami Council and the U.S.S.R. Association of Small Peoples of the North. The Meetings on the Arctic Environment shall serve to:

- Identify and coordinate actions to implement and further develop the Arctic Environmental Protection Strategy;
- Initiate cooperation in new fields relevant to the environmental protection of the Arctic;
- Make necessary recommendations in order to protect the Arctic environment;
- Improve existing environmental regimes relevant to the Arctic;
- Assess and report on progress on actions agreed upon.

Chapter 3

The Environment Act

PRELIMINARY

SHORT TITLE, EXTEND AND COMMENCEMENT

- This Act may be called the Environment (Protection) Act, 1986.
- It extends to the whole of India.
- It shall come into force on such date as the Central Government may, by notification in the Official Gazette, appoint and different dates may be appointed for different provisions of this Act and for different areas.

DEFINITIONS

In this Act, unless the context otherwise requires::

- "Environment" includes water, air and land and the inter-relationship which exists among and between water, air and land, and human beings, other living creatures, plants, micro-organism and property;
- "Environmental pollutant" means any solid, liquid or gaseous substance present in such concentration as may be, or tend to be, injurious to environment;
- "Environmental pollution" means the presence in the environment of any environmental pollutant;
- "Handling", in relation to any substance, means the manufacture, processing, treatment, package, storage, transportation, use, collection, destruction, conversion, offering for sale, transfer or the like of such substance;
- "Hazardous substance" means any substance or preparation which, by reason of its chemical or physico-chemical properties or handling, is liable to cause harm to human beings, other living creatures, plant, micro-organism, property or the environment;
- "Occupier", in relation to any factory or premises, means a person who has, control over the affairs of the factory or the

premises and includes in relation to any substance, the person in possession of the substance;

- "Prescribed" means prescribed by rules made under this Act.

GENERAL POWERS OF THE CENTRAL GOVERNMENT

POWER OF CENTRAL GOVERNMENT TO TAKE MEASURES TO PROTECT AND IMPROVE ENVIRONMENT

- Subject to the provisions of this Act, the Central Government, shall have the power to take all such measures as it deems necessary or expedient for the purpose of protecting and improving the quality of the environment and preventing controlling and abating environmental pollution.
- In particular, and without prejudice to the generality of the provisions of sub-section (1), such measures may include measures with respect to all or any of the following matters, namely:
- Co-ordination of actions by the State Governments, officers and other authorities:
  - Under this Act, or the rules made thereunder, or
  - Under any other law for the time being in force which is relatable to the objects of this Act;
- Planning and execution of a nation-wide programme for the prevention, control and abatement of environmental pollution;
- Laying down standards for the quality of environment in its various aspects;
- Laying down standards for emission or discharge of environmental pollutants from various sources whatsoever:

Provided that different standards for emission or discharge may be laid down under this clause from different sources having regard to the quality or composition of the emission or discharge of environmental pollutants from such sources;

- Restriction of areas in which any industries, operations or processes or class of industries, operations or processes shall not be carried out or shall be carried out subject to certain safeguards;
- Laying down procedures and safeguards for the prevention of accidents which may cause environmental pollution and remedial measures for such accidents;
- Laying down procedures and safeguards for the handling of hazardous substances;
- Examination of such manufacturing processes, materials and substances as are likely to cause environmental pollution;
- Carrying out and sponsoring investigations and research relating to problems of environmental pollution;

- Inspection of any premises, plant, equipment, machinery, manufacturing or other processes, materials or substances and giving, by order, of such directions to such authorities, officers or persons as it may consider necessary to take steps for the prevention, control and abatement of environmental pollution;
- Establishment or recognition of environmental laboratories and institutes to carry out the functions entrusted to such environmental laboratories and institutes under this Act;
- Collection and dissemination of information in respect of matters relating to environmental pollution;
- Preparation of manuals, codes or guides relating to the prevention, control and abatement of environmental pollution;
- Such other matters as the Central Government deems necessary or expedient for the purpose of securing the effective implementation of the provisions of this Act.
- The Central Government may, if it considers it necessary or expedient so to do for the purpose of this Act, by order, published in the Official Gazette, constitute an authority or authorities by such name or names as may be specified in the order for the purpose of exercising and performing such of the powers and functions (including the power to issue directions under section (5) of the Central Government under this Act and for taking measures with respect to such of the matters referred to in sub-section (2) as may be mentioned in the order and subject to the supervision and control of the Central Government and the provisions of such order, such authority or authorities may exercise and powers or perform the functions or take the measures so mentioned in the order as if such authority or authorities had been empowered by this Act to exercise those powers or perform those functions or take such measures.

APPOINTMENT OF OFFICERS AND THEIR POWERS AND FUNCTIONS

- Without prejudice to the provisions of sub-section (3) of section 3, the Central Government may appoint officers with such designation as it thinks fit for the purposes of this Act and may entrust to them such of the powers and functions under this Act as it may deem fit.
- The officers appointed under sub-section (1) shall be subject to the general control and direction of the Central Government or, if so directed by that Government, also of the authority or

authorities, if any, constituted under sub- section (3) of section 3 or of any other authority or officer.

POWER TO GIVE DIRECTIONS

Notwithstanding anything contained in any other law but subject to the provisions of this Act, the Central Government may, in the exercise of its powers and performance of its functions under this Act, issue directions in writing to any person, officer or any authority and such person, officer or authority shall be bound to comply with such directions.

Explanation: For the avoidance of doubts, it is hereby declared that the power to issue directions under this section includes the power to direct.

- The closure, prohibition or regulation of any industry, operation or process.
- Stoppage or regulation of the supply of electricity or water or any other service.

RULES TO REGULATE ENVIRONMENTAL POLLUTION

- The Central Government may, by notification in the Official Gazette, make rules in respect of all or any of the matters referred to in section 3.
- In particular, and without prejudice to the generality of the foregoing power, such rules may provide for all or any of the following matters, namely:
  - The standards of quality of air, water or soil for various areas and purposes;
  - The maximum allowable limits of concentration of various environmental pollutants (including noise) for different areas;
  - The procedures and safeguards for the handling of hazardous substances;
  - The prohibition and restrictions on the handling of hazardous substances in different areas;
  - The prohibition and restriction on the location of industries and the carrying on process and operations in different areas;
  - The procedures and safeguards for the prevention of accidents which may cause environmental pollution and for providing for remedial measures for such accidents.

PREVENTION, CONTROL, AND ABATEMENT OF ENVIRONMENTAL POLLUTION

PERSONS CARRYING ON INDUSTRY OPERATION

No person carrying on any industry, operation or process shall discharge or emit or permit to be discharged or emitted any environmental pollutants in excess of such standards as may be prescribed.

PERSONS HANDLING HAZARDOUS SUBSTANCES TO COMPLY WITH PROCEDURAL SAFEGUARDS

No person shall handle or cause to be handled any hazardous substance except in accordance with such procedure and after complying with such safeguards as may be prescribed.

FURNISHING OF INFORMATION TO AUTHORITIES AND AGENCIES IN CERTAIN CASES

- Where the discharge of any environmental pollutant in excess of the prescribed standards occurs or is apprehended to occur due to any accident or other unforeseen act or event, the person responsible for such discharge and the person in charge of the place at which such discharge occurs or is apprehended to occur shall be bound to prevent or mitigate the environmental pollution caused as a result of such discharge and shall also forthwith:
  - Intimate the fact of such occurrence or apprehension of such occurrence.
  - Be bound, if called upon, to render all assistance, to such authorities or agencies as may be prescribed.
- On receipt of information with respect to the fact or apprehension on any occurrence of the nature referred to in sub-section (1), whether through intimation under that sub-section or otherwise, the authorities or agencies referred to in sub-section (1) shall, as early as practicable, cause such remedial measures to be taken as necessary to prevent or mitigate the environmental pollution.
- The expenses, if any, incurred by any authority or agency with respect to the remedial measures referred to in sub-section (2), together with interest (at such reasonable rate as the Government may, by order, fix) from the date when a demand for the expenses is made until it is paid, may be recovered by such authority or agency from the person concerned as arrears of land revenue or of public demand.

POWERS OF ENTRY AND INSPECTION

- Subject to the provisions of this section, any person empowered by the Central Government in this behalf shall have a right to enter, at all reasonable times with such assistance as he considers necessary, any place:

- For the purpose of performing any of the functions of the Central Government entrusted to him.
- For the purpose of determining whether and if so in what manner, any such functions are to be performed or whether any provisions of this Act or the rules made thereunder orany notice, order, direction or authorisation served, made, given or granted under this Act is being or has been complied with.
- For the purpose of examining and testing any equipment, industrial plant, record, register, document or any other material object or for conducting a search of any building in which he has reason to believe that an offence under this Act or the rules made thereunder has been or is being or is about to be committed and for seizing any such equipment, industrial plant, record, register, document or other material object if he has reason to believe that it may furnish evidence of the commission of an offence punishable under this Act or the rules made thereunder or that such seizure is necessary to prevent or mitigate environmental pollution.

- Every person carrying on any industry, operation or process of handling any hazardous substance shall be bound to render all assistance to the person empowered by the Central Government under sub-section (1) for carrying out the functions under that sub-section and if he fails to do so without any reasonable cause or excuse, he shall be guilty of an offence under this Act.
- If any person wilfully delays or obstructs any persons empowered by the Central Government under sub-section (1) in the performance of his functions, he shall be guilty of an offence under this Act.
- The provisions of the Code of Criminal Procedure, 1973, or, in relation to the State of Jammu and Kashmir, or an area in which that Code is not in force, thc provisions of any corresponding law in force in that State or area shall, so far as may be, apply to any search or seizures under this section as they apply to any search or seizure made under the authority of a warrant issued under section 94 of the said Code or as the case may be, under the corresponding provision of the said law.

POWER TO TAKE SAMPLE AND PROCEDURE TO BE FOLLOWED IN CONNECTION THEREWITH

- The Central Government or any officer empowered by it in this behalf, shall have power to take, for the purpose of analysis, samples of air, water, soil or other substance from any factory, premises or other place in such manner as may be prescribed.

- The result of any analysis of a sample taken under sub-section (1) shall not be admissible in evidence in any legal proceeding unless the provisions of sub-sections (3) and (4) are complied with.
- Subject to the provisions of sub-section (4), the person taking the sample under sub-section (1) shall:
  - Serve on the occupier or his agent or person in charge of the place, a notice, then and there, in such form as may be prescribed, of his intention to have it so analysed;
  - In the presence of the occupier of his agent or person, collect a sample for analysis;
  - Cause the sample to be placed in a container or containers which shall be marked and sealed and shall also be signed both by the person taking the sample and the occupier or his agent or person;
  - Send without delay, the container or the containers to the laboratory established or recognised by the Central Government under section 12.
- When a sample is taken for analysis under sub-section (1) and the person taking the sample serves on the occupier or his agent or person, a notice under clause (a) of sub-section (3), then,—
  - In a case where the occupier, his agent or person wilfully absents himself, the person taking the sample shall collect the sample for analysis to be placed in a container or containers which shall be marked and sealed and shall also be signed by the person taking the sample, and
  - In a case where the occupier or his agent or person present at the time of taking the sample refuses to sign the marked and sealed container or containers of the sample as required under clause (c) of sub-section (3), the marked and sealed container or containers shall be signed by the person taking the samples, and the container or containers shall be sent without delay by the person taking the sample for analysis to the laboratory established or recognised under section 12 and such person shall inform the Government Analyst appointed or recognised under section 12 in writing, about the wilfull absence of the occupier or his agent or person, or, as the case may be, his refusal to sign the container or containers.

ENVIRONMENTAL LABORATORIES

- The Central Government may, by notification in the Official Gazette:

- Establish one or more environmental laboratories.
- Recognise one or more laboratories or institutes as environmental laboratories to carry out the functions entrusted to an environmental laboratory under this Act.

- *The Central Government may, by notification in the Official Gazette, make rules specifying:*
  - The functions of the environmental laboratory.
  - The procedure for the submission to the said laboratory of samples of air, water, soil or other substance for analysis or tests, the form of the laboratory report thereon and the fees payable for such report;
  - Such other matters as may be necessary or expedient to enable that laboratory to carry out its functions.

GOVERNMENT ANALYSTS

The Central Government may by notification in the Official Gazette, appoint or recognise such persons as it thinks fit and having the prescribed qualifications to be Government Analysts for the purpose of analysis of samples of air, water, soil or other substance sent for analysis to any environmental laboratory established or recognised under sub-section (1) of section 12.

REPORTS OF GOVERNMENT ANALYSTS

Any document purporting to be a report signed by a Government analyst may be used as evidence of the facts stated therein in any proceeding under this Act.

PENALTY FOR CONTRAVENTION OF THE PROVISIONS OF THE ACT AND THE RULES, ORDERS AND DIRECTIONS

- Whoever fails to comply with or contravenes any of the provisions of this Act, or the rules made or orders or directions issued thereunder, shall, in respect of each such failure or contravention, be punishable with imprisonment for a term which may extend to five years with fine which may extend to one lakh rupees, or with both, and in case the failure or contravention continues, with additional fine which may extend to five thousand rupees for every day during which such failure or contravention continues after the conviction for the first such failure or contravention.
- If the failure or contravention referred to in sub-section (1) continues beyond a period of one year after the date of conviction, the offender shall be punishable with imprisonment for a term which may extend to seven years.

OFFENCES BY COMPANIES

- Where any offence under this Act has been committed by a company, every person who, at the time the offence was committed, was directly in charge of, and was responsible to, the company for the conduct of the business of the company, as well as the company, shall be deemed to be guilty of the offence and shall be liable to be proceeded against and punished accordingly.

Provided that nothing contained in this sub-section shall render any such person liable to any punishment provided in this Act, if he proves that the offence was committed without his knowledge or that he exercised all due diligence to prevent the commission of such offence.

- Notwithstanding anything contained in sub-section (1), where an offence under this Act has been committed by a company and it is proved that the offence has been committed with the consent or connivance of, or is attributable to any neglect on the part of, any director, manager, secretary or other officer of the company, such director, manager, secretary or other officer shall also deemed to be guilty of that offence and shall be liable to be proceeded against and punished accordingly.

Explanation:

- For the purpose of this section.
- "Company" means any body corporate and includes a firm or other association of individuals.
- "Director", in relation to a firm, means a partner in the firm.

OFFENCES BY GOVERNMENT DEPARTMENTS

- Where an offence under this Act has been committed by any Department of Government, the Head of the Department shall be deemed to be guilty of the offence and shall be liable to be proceeded against and punished accordingly.

Provided that nothing contained in this section shall render such Head of the Department liable to any punishment if he proves that the offence was committed without his knowledge or that he exercise all due diligence to prevent the commission of such offence.

- Notwithstanding anything contained in sub-section (1), where an offence under this Act has been committed by a Department of Government and it is proved that the offence has been committed with the consent or connivance of, or is attributable to any neglect on the part of, any officer, other than the Head of the Department, such officer shall also be deemed to be guilty of that offence and shall be liable to be proceeded against and punished accordingly.

MISCELLANEOUS

PROTECTION OF ACTION TAKEN IN GOOD FAITH

No suit, prosecution or other legal proceeding shall lie against the Government or any officer or other employee of the Government or any authority constituted under this Act or any member, officer or other employee of such authority in respect of anything which is done or intended to be done in good faith in pursuance of this Act or the rules made or orders or directions issued thereunder.

COGNIZANCE OF OFFENCES

No court shall take cognizance of any offence under this Act except on a complaint made by:

- The Central Government or any authority or officer authorised in this behalf by that Government.
- Any person who has given notice of not less than sixty days, in the manner prescribed, of the alleged offence and of his intention to make a complaint, to the Central Government or the authority or officer authorised as aforesaid.

INFORMATION, REPORTS OR RETURNS

The Central Government may, in relation to its function under this Act, from time to time, require any person, officer, State Government or other authority to furnish to it or any prescribed authority or officer any reports, returns, statistics, accounts and other information and such person, officer, State Government or other authority shall be bound to do so.

MEMBERS, OFFICERS AND EMPLOYEES OF THE AUTHORITY CONSTITUTED UNDER SECTION 3 TO BE PUBLIC SERVANTS

All the members of the authority, constituted, if any, under section 3 and all officers and other employees of such authority when acting or purporting to act in pursuance of any provisions of this Act or the rules made or orders or directions issued thereunder shall be deemed to be public servants within the meaning of section 21 of the Indian Penal Code (45 of 1860).

BAR OF JURISDICTION

No civil court shall have jurisdiction to entertain any suit or proceeding in respect of anything done, action taken or order or direction issued by the Central Government or any other authority or officer in pursuance of any power conferred by or in relation to its or his functions under this Act.

POWERS TO DELEGATE

Without prejudice to the provisions of sub-section (3) of section 3, the Central Government may, by notification in the Official Gazette, delegate, subject to such conditions and limitations as may be specified in the notifications, such of its powers and functions under this Act [except the powers to constitute an authority under sub-section (3) of section 3 and to make rules under section 25] as it may deem necessary or expedient, to any officer, State Government or other authority.

EFFECT OF OTHER LAWS

- Subject to the provisions of sub-section (2), the provisions of this Act and the rules or orders made therein shall have effect notwithstanding anything inconsistent therewith contained in any enactment other than this Act.
- Where any act or omission constitutes an offence punishable under this Act and also under any other Act then the offender found guilty of such offence shall be liable to be punished under the other Act and not under this Act.

POWER TO MAKE RULES

- The Central Government may, by notification in the Official Gazette, make rules for carrying out the purposes of this Act.
- In particular, and without prejudice to the generality of the foregoing power, such rules may provide for all or any of the following matters, namely:
  - The standards in excess of which environmental pollutants shall not be discharged or emitted under section 7.
  - The procedure in accordance with and the safeguards in compliance with which hazardous substances shall be handled or caused to be handled under section 8.
  - The authorities or agencies to which intimation of the fact of occurrence or apprehension of occurrence of the discharge of any environmental pollutant in excess of the prescribed standards shall be given and to whom all assistance shall be bound to be rendered under sub-section (1) of section 9.
  - The manner in which samples of air, water, soil or other substance for the purpose of analysis shall be taken under sub-section (1) of section 11.
  - The form in which notice of intention to have a sample analysed shall be served under clause (a) of sub section (3) of section 11;
  - The functions of the environmental laboratories, the

procedure for the submission to such laboratories of samples of air, water, soil and other substances for analysis or test; the form of laboratory report; the fees payable for such report and other matters to enable such laboratories to carry out their functions under sub-section (2) of section 12.

- The qualifications of Government Analyst appointed or recognised for the purpose of analysis of samples of air, water, soil or other substances under section 13.
- The manner in which notice of the offence and of the intention to make a complaint to the Central Government shall be given under clause (b) of section 19.
- The authority of officer to whom any reports, returns, statistics, accounts and other information shall be furnished under section 20.
- Any other matter which is required to be, or may be, prescribed.

RULES MADE UNDER THIS ACT TO BE LAID BEFORE PARLIAMENT

Every rule made under this Act shall be laid, as soon as may be after it is made, before each Hose of Parliament, while it is in session, for a total period of thirty days which may be comprised in one session or in two or more successive sessions, and if, before the expiry of the session immediately following the session or the successive sessions aforesaid, both Houses agree in making any modification in the rule or both Houses agree that the rule should not be made, the rule shall thereafter have effect only in such modified form or be of no effect, as the case may be; so, however, that any such modification or annulment shall be without prejudice to the validity of anything previously done under that rule.

Chapter 4

Integrated Management for Environmental Protection

WATER QUALITIY

Water is a scarce commodity in Jordan. The per capita share of indigenous renewable freshwater resources has been estimated at 175 m3 in 1996. A mere 15C liters per capita per day is all that can the country afford to allocate municipal water to its citizens at the present time. These figures are amongst the lowest in the world and even in the Middle Eastern region.

The country is heading towards reaching the unsafe edges of water affordability and accessibility. Demand will outstrip the total exploitable resources by the year 2000; 1095 and 869 million cubic meters per year respectively. Water scarcity is not only associated with the quantity available but also with the quality which should meet specified standards in order to be useful.

The water scarcity problem is further growth rate estimated at 3.5% (a combined natural increase plus immigration). Although it is expected to decrease to 3.3% by 2005, it remains amongst the highest rates in the world. With increasing populations, Jordan will be faced with the dual dilemma of increasing demands, on the already limited freshwater resources, and more wastes generated. Thus, freshwater extraction will be on the increase, as will threats to the quality of these water resources.

Hence, the protection and management of freshwater resources must focus on protecting and managing both their quantity and their quality. Transboundary issues are becoming important. High proportions of the water resources currently available to Jordan are resources shared with neighboring countries. The quality of such resources can be highly dependent on the behaviour of these riparian partners. Jordan should ensure that the quality of such resources is protected through the implementation of water quality monitoring programs. The following Programme Areas are proposed in order to protect the quality of Jordan water resources.

REGULATORY AND STANDARDS ISSUES

Basis for Action

Regulatory standards are based on guidelines issued by the World Health Organization (WHO), the Food and Agriculture Organization of the United Nations (FAO) and other international standards including those enacted in Europe and the United States. This approach to national standards was perhaps inevitable in the short-term, but it is widely recognized to be inappropriate for the longer-term.

For the longer-term Jordan needs to promulgate national standards adapted to meet local conditions Only six of the 16 wastewater treatment plants in Jordan were reported in 1996 to have met their design performance expectations. These six were also the only ones producing effluents conforming to the requirements of the Jordanian Regulatory Standard JS: 893/1995 for discharge to wadis. Upgrading the effluent of the other 10 WWTP is necessary to protect both surface and groundwater resources.

MONITORING PROGRAMS AND DATA MANAGEMENT

Basis for Action.

Several governmental and quasigovernmental departments with little or no co-ordination among them perform water quality monitoring in Jordan. Monitoring programs are primarily carried out at the laboratories of the Ministry of Water and Irrigation, Ministry of Health and the Royal Scientific Society (RSS).

The latter has been subcontracted by the General Corporation for Environmental Protection (GCEP). However, lack of co-ordination among these departments has led to inefficient utilization of the national testing capabilities; duplication of efforts among these agencies and dilution of responsibility.

INNOVATIVE TECHNOLOGY AND RESEARCH

Basis for Action

Little, if any, applied original research in the water quality field, has been carried out to deal with water issues specific to Jordan. There are several potential areas of water quality research that can be only properly achieved through national initiatives and efforts. Jordan should strengthen and encourage local water quality research. Research on water salinity, production of locally tailored water quality standards for various uses, and conducting epidemiological studies are particularly needed.

PUBLIC AWARENESS AND WATER QUALITY EDUCATION

Basis for Action

Public awareness of the extent and severity of the water scarcity problem Jordan is facing has been modest, and so have been the attitudes in dealing with it. Public knowledge of water issues, whether related to water utilization or water protection is insufficient in light of the seriousness of national water issues.

This can be attributed to complacency on the public side and to a lack of communication between planners and users. It is only in the last few years that water conservation awareness campaigns have been launched in the media. Most of these awareness programs concentrate on the urgency of using less water by the general public. However, governmental institutions, administrators, private sector business and industrial people, academics, religious groups and the mass media at large ought to be targeted as well.

Such campaigns should also reflect the integrated nature of water plans. Issues that should be particularly targeted in these campaigns include: the consequences of water quality deterioration, the need to protect scarce freshwater resources from pollution and the need to refrain from using high quality water for purposes that can be satisfied with lower quality water.

WASTEWATER MANAGEMENT

MUNICIPAL WASTEWATER

The environmental and health problems associated with human liquid wastes have changed over the years. The cities, towns, and villages of Jordan spilled out uncontrollably into the surrounding agricultural land under the pressures of population movements, which have been precipitated by the political and military events since 1948, as well as a high rate of natural growth.

Traditional methods of individual on-site excreta disposal were no longer suitable for congested population centres, in part due to their impact on local ground water aquifers. In 1968, the Ein Ghazal Activated Sludge(AS) plant was commissioned for Amman. The plant was designed to handle a hydraulic load of 260,000 m^3/day and an organic load of 18 tons/day BOD5, based on a design population of 300,000 and water use of 200 l/c/d and wastewater strength of 300 mg l. A cholera epidemic in 1980 paved the road for abandoning the Ein Ghazal plant and construction of the Khirbet Al-Samra Stabilization ponds. The epidemic highlighted, at once, the threats inherent in uncontrolled re-use of properly treated wastewater for irrigation under adequate environmental health safeguards, and the problem of septage.

The 1980s witnessed the introduction of modern sewerage systems into the main cities of Jordan, with treatment plants replacing the traditional individual cesspits then in common use. By 1997, there were about 16 municipal Wastewater Treatment Plants (WWT) serving about 2.4 million people (57% of the total), 16 smaller plants for institutions (hospitals, etc.) and 3 for industrial locations.

EXTENDING SEWERAGE SERVICE COVERAGE

Basis for Action

During the last two decades, considerable progress has been achieved in the wastewater sector. -1990, contributed to this. This progress was also a result of intensive programs constructing sewerage collection and treatment plants. By 1997, about 16 municipal treatment plants were in operation, including the Samra stabilization ponds, receiving about 70% of total wastewater treated. However, despite these achievements, there are a number of issues that remain to be tackled, including the following:

- Public sewer networks reach about 80% of households in Greater Amman, but only 4.5% in other urban areas, making a countrywide average of 25% households (about 50% of the population).
- There is still room for expansion of the Greater Amman system and others and need for establishment of new systems for the many un-sewered urban centres, urban fringes and slum impoverished settlements.
- Around 12% of households nationwide, mostly in rural areas are Moreover, there is a need to improve the design, construction and maintenance of the individual on-site disposal systems; septage removal and disposal is also another issue in the sector.
- Many cases of pollution of water and food occur annually, many of these are caused by intrusion of sewage into water systems, through weak joints or broken sewers, blockages and overflows.
- Strength of raw influent entering treatment plants is in the range

of 500-1600 BOD5 mg/l. This imposes several operational problems; even with high removal efficiency, the effluent BOD5 will still have a BOD5 high enough to cause re-use and disposal problems.

PROMOTING INTEGRATED WASTEWATER MANAGEMENT

Basis for Action

The wastewater sector has received due attention and support since the early eighties. Since then, the government has achieved a high level of wastewater collection, treatment and re-use. However, there has been a lag in management aspects of this sector, which has been overshadowed all the time by the water supply sector.

Despite the achievements of the Ministry of Water and Irrigation in the construction of sewerage collection and treatment systems, some crucial issues remain to be resolved, including: There is no national wastewater strategy with comprehensive goals and policies. The institutional structure suffers from the lack of clear lines of authority and responsibilities, and from weak coordination and integration; all these factors can weaken the delivery of services.

There is no clear and precise definition and description of the duties and tasks of different administrative units. Management of personnel suffers from deficiencies including vague or non-existent job descriptions, lack of incentives and motivation, lack of systematic training activities for developing and enhancing the performance and output of staff.

There are deficiencies in the operation and maintenance of the wastewater collection system (house connections, sewers, appurtenances, etc.) and the treatment plants (structures, equipment, treatment operations, records, etc.). Breakdowns in the sewerage system and the treatment plants occur occasionally; failure to respond to these breakdowns adequately and promptly create hazardous situations and precipitates public health risks.

MAXIMIZING ENVIRONMENTALLY SOUND REUSE AND RECYCLING

Basis for Action

Water demand in Jordan cannot be met by the available supply capacity. In the semi-arid conditions of Jordan, reuse of wastewater is a necessity. It is estimated that in 1997 about 73.5 MCM of wastewater were received in the 16 municipal treatment plants, of which about

61.7 MCM (83 %) were used directly or indirectly for irrigation. This represents about 10% of total water available for irrigation. (The irrigated area in Jordan is estimated at: 14,000 dunums for controlled irrigation and 81,000 dunums for unrestricted irrigation).

A number of issues that have to be considered: More than 70% of all wastewater is treated in one facility, the Khirbet Al Samra waste stabilization ponds (WSP). Several treatment plants are overloaded, most notably the Khirbet Al Samra where loading of the ponds reaches about twice the designed capacity. Most re-use is indirect; effluent is discharged to a natural wadi (seil) and withdrawn some distance downstream along the wadi The Al Samra facility discharges its effluent into the Zarqa river and ultimately into the King Talal Dam.

PROMOTING ENVIRONMENTALLY SOUND WASTEWATER DISPOSAL AND TREATMENT

Basis for Action

Waste minimization is a basic tool in advancing sustainability because it helps reduce the amounts of all types of wastes released into the environment. However, the released wastes still need to be treated and disposed of by safe, environmentally sound systems. Jordan has achieved a high level of sewerage treatment. In 1997, municipal systems treated about 73.5 MCM of wastewater in 16 treatment plants, serving about 2.4 million population.

However, there are issues that require attention and consideration: Many communities still depend on cesspools for sewage disposal, but such systems have become inadequate due to the increase in population, limited lots of land and unsuitable geological/topographical conditions. In rural areas, where individual systems exist, about 3% of population lack proper access to appropriate sanitation services.

Practices in unplanned land use have encroached on fertile lands, scattering the housing schemes, industrial estates and other developmental projects causing problems of provision of adequate services and control of health Residential

and sanitary impacts. Effluents from WWT plants are discharged into the nearest seil (wadi) which is usually dry for most of the year.

Some effluent seeps into the groundwater and a considerable portion of it is used for irrigation, though the quality of the treated effluent is not acceptable for unrestricted irrigation. Due to unforeseen pressures, such as population movements and financial constraints, the pace of providing sewerage systems has slowed down; treatment plants are overloaded.

INDUSTRIAL WASTEWATER

Industrialization involves major activities promoted by the Government of Jordan in its development policies and strategies. Industry is becoming increasingly important in social and economic development and a major contributor to the Gross Domestic Product. In addition, industry is providing employment and livelihood opportunities and contributing to the enhancement of human welfare in Jordan. Industry plays a key role in helping to achieve sustainable development through more efficient production processes, preventive strategies and cleaner production technologies and procedures throughout the life cycle of the product, hence minimizing or avoiding wastes. Industries should be full participants in the implementation of activities related to Agenda 21.

Environmental management should be among the highest priorities and a key element to sustainable development. This can be achieved by taking into consideration environmental criteria for the siting of industries and adopting guidelines for assessing environmental impacts of new industries, carrying out audits and assessments of compliance and openness with the public. The improvement of production systems through technologies and processes that utilize resources more efficiently and at the same time produce less wastes - achieving more with less - is another important pathway towards sustainability for industry.

POLLUTION PREVENTION AND CONTROL

Basis for Action

Control is often expensive and requires complex regulatory efforts. By reducing pollutants before they enter the environment, the risk to human health and the environment is decreased, and the cost of regulating and cleaning up pollutants is less. This implies moving towards pollution prevention and cleaner production, optimizing efficiencies at each stage of production. Cleaner production can improve the overall competitiveness of an enterprise, improving its efficiency of resource use, minimizing waste and protecting human health and environmental quality.

ENVIRONMENTAL MANAGEMENT

Basis for Action

There are two main features of Environmental Management that industries should adopt: Environmental Impact Assessment (EIA) and

Environmental Auditing. Siting of an industry will have an important effect on its subsequent environmental impact. It is acknowledged that industrial activities produce physical and social changes. Ignoring this fact in Jordan has caused several industries to be sited where they caused adverse environmental impacts on the surrounding environment, especially water resources.

EIA is a technique that can be used to integrate environmental considerations into industrial projects. EIA should be carried out early in the planning and feasibility stage so that alternative processes, raw materials, siting options and courses of action can be evaluated and environmental impacts identified and mitigated. Auditing can be broadly defined as, a systematic evaluation of a environment, safety and health. Auditing can be viewed as a voluntarily approach that benefit industry and the environment and has the advantage of being a proactive preventive approach rather than a reactive approach to legislation. Industries in Jordan should develop techniques to reduce harmful environmental impacts through audits and assessment of compliance with environmental laws and regulations

MUNICIPAL SOLID WASTE MANAGEMENT

Jordan has experienced an increase in the volume of solid waste generation in the last decade. This was the result of the population pressure, industrial development, and new consumption patterns and lifestyles. Solid waste management is the responsibility of local government and common services councils in the country. The capacities of these authorities to deal with solid waste issues are hindered by the lack of financial resources, lack of skilled manpower and lack of community awareness.

MINIMIZING GENERATION OF SOLID WASTES

Basis for Action

The level of municipal solid wastes (MSW) generated in Jordan is relatively high at 0.9 kg per capita per day. MSW are expected to increase with time due to the high annual population growth rate of 3.5% and due to unsustainable patterns of production and consumption. This imposes a heavy burden on the existing solid waste disposal sites. Most of these sites will be exhausted within the next few years, including the Russifeh landfill site which currently serves 2,250,000 inhabitants. Selection of new disposal sites and managing them in an environmentally sound manner will increase the cost of MSW handling and disposal. Thus, the first step in MSW management is to focus on reducing the generation of wastes destined for final disposal by attempting to change unsustainable patterns of production and consumption.

MAXIMIZING ENVIRONMENTALLY SOUND SOLID WASTE REUSE AND RECYCLING

Basis for Action

As mentioned earlier, it is expected that the cost of MSW disposal will increase due to emerging stringent environmental regulations and the exhaustion of current disposal sites. MSW management will need to promote reuse and recycling to help minimize waste generation and reduce wastes to be consigned to disposal sites. MSW in Jordan includes a mix of domestic garbage, plastics, glass, metals in addition to hazardous and non-hazardous medical and industrial solid wastes. Sorting of the different types of solid waste at generation sources is essential to facilitate recycling and reuse. This is not yet practiced in Jordan, and considerable amounts of recyclable materials are sent to final disposal sites. Scavengers collect aluminum soft drinks cans and other metals mainly at final disposal sites. Their work is inefficient and imposes adverse impact on their health

PROMOTING ENVIRONMENTALLY SOUND SOLID WASTE DISPOSAL AND TREATMENT

Basis for Action

There are about 30 solid waste disposal sites in Jordan of which only 10 are authorized. Environmental concerns are not related only to unauthorized sites, but to authorized sites as well. This is due to the fact that no environmental impact assessments were conducted prior to developing these sites. The method of disposal in most sites is open dumping. Some disposal sites receive medical and/or industrial hazardous solid waste, sludge from wastewater treatment plants and/or liquid industrial waste. In conclusion, design and operation of most (if not all) disposal sites are not in accordance with environmentally sound practices. Environmental impacts of disposal sites on groundwater, air quality and other environment components raise serious public health concerns.

EXTENDING WASTE SERVICE COVERAGE

Basis for Action

Solid waste services are available for population, while the percentage of served population as a nationwide average is approximately 75%. There is a real challenge to extend the services to all people taking into consideration the current coverage percentages and the high growth rate Safe solid waste collection and disposal services are crucial for public health protection. Without these services public health is endangered as a result of air, water and soil pollution, and waste-related diseases.

MANAGEMENT OF HAZARDOUS WASTES

Understanding the negative impacts that some chemicals will have on human life and the extent to which these chemicals will effect the environment is becoming a vital issue to many Jordanians. An increased awareness of the importance of the effective management of hazardous wastes, and its role in environmental and public health protection and sustainable development, is steadily mounting. The overall objective of the following programme areas is to minimize the generation of hazardous wastes as much as possible and to manage these chemicals

STRENGTHENING INSTITUTIONAL CAPACITIES FOR HAZARDOUS WASTE MANAGEMENT

Basis for Action

There is a general lack of data on the quantities and types of hazardous materials that have been disposed of in the environment and the location where they have been dumped. Hazardous materials are not properly disposed of in licensed disposal sites (which are nonexistent) and there are no efforts to estimate the extent of contamination caused by these materials or to propose clean up schemes for the contaminated sites.

There is no agency able to provide the public or specialists with technical information on hazardous chemicals. Comprehensive regulations or standards related to hazardous chemicals are not available. Despite the establishment of the General Corporation for the Environment Protection (GCEP) and the law for the environmental protection (Law No.12 of 1995), the responsibility for implementing several environmental regulations by GCEP interfered with the responsibilities of other governmental agencies or ministries.

Wastes generated from hospitals, laboratories and research centres, such as radioactive materials, chemical solvents, paints, cleaning agents, hazardous materials from local factories and small size industries, are mixed in with the domestic solid wastes, despite the fact that there is no leachate collection systems or monitoring programs for these sites.

Current practices designed to treat and dispose of hazardous wastes are widely violated.

Monitoring programs are needed to determine the extent of contamination by hazardous chemicals and its effect on public health. There are shortages of sufficiently trained personnel in the areas related to hazardous substances handling, transporting, treatment and disposing of. No substantial research is conducted in this field in the Jordanian institutions. There is a need for an emergency plan in case of hazardous chemical spills or accidents. There is a lack of awareness programs to educate the public on hazardous wastes.

PROMOTING MINIMIZATION AND PREVENTION OF HAZARDOUS WASTE

Basis for Action

Management of hazardous substances is costly, and thus minimization and prevention of hazardous wastes must have the first priority. With the current technological and economical resources in Jordan, management of hazardous wastes in an environmentally sound matter is not possible. No hazardous waste landfill is yet available in Jordan. There is a general lack of knowledge and information on the economics of prevention and management of hazardous wastes. There have been no substantial efforts to change the industrial processes and consumption patterns through pollution prevention and cleaner production strategies.

Basis for Action

There is a need for information exchange on the benefits and risks associated with the use of chemicals, in order to enhance the sound management of hazardous chemicals through the exchange of scientific, technical, economic and legal information. The import of some hazardous substances that have been banned in producing countries has been an area of concern. Available information is inadequate for insuring the safe use, control, distribution, storage, formulation and disposal of restricted chemicals.

PREVENTING ILLEGAL TRAFFIC OF HAZARDOUS WASTES

Basis for Action

There is a need to harmonize existing criteria for identifying hazardous wastes and the procedures used in various international and legal instruments. There is a lack of monitoring systems in Jordan. There is also a lack of awareness programs to promote the use of a precautionary approach to the control and monitoring of transboundary movements of hazardous chemicals.

SAFE AND ENVIRONMENTALLY SOUND MANAGEMENT OF RADIOACTIVE WASTES

Basis for Action

Nuclear applications in Jordan span the fields of medicine, research, agriculture, water, industry, mineralogy, etc. Radioactive wastes although produced in small quantities from these applications, need stringent disposal measures. Plans are underway to start a Jordanian nuclear programme that will include the construction of a nuclear reactor. There is a potential for high levels of radiation due to the nuclear programs in the neighboring countries, especially in Israel, where nuclear accidents are anticipated.

Nuclear accidents are very likely to occur in the nuclear reactors or nuclear programs in the neighboring countries, especially Israel. There is a need for a national emergency plan in case of nuclear or radiation accidents. Monitoring programs are needed to determine the radiation levels all over the kingdom, especially in areas adjacent to the Israeli nuclear reactors, as well as a comprehensive survey to identify areas with non-ionizing radiation levels detrimental to public health.

There is a shortage of welltrained personnel in the areas related to radioactive substances handling, transporting, and management in general. There is a lack of detailed specifications and standards to regulate the radiation levels from electrical and electronic devices locally made and radiation levels from wastes in the work environment. No substantial research is conducted in this field in Jordanian institutions. There is a lack of awareness programs to educate the public on handling radioactive waste or responding to radiation accidents.

AIR QUALITY CONTROL

During the period between 1980 and 1997, the emission of five principle pollutants increased significantly in Jordan. These five pollutants are: particulate matter (TSP and PM10), sulfur dioxide (SO_2), carbon monoxide (CO), nitrogen dioxide (NO_2), and lead (Pb). Article (18) of the Environmental Law No.12/1995, provides the framework for pertinent organizations to protect air quality. The article also describes the rcle of the GCEP in the field of ai: protection. GCEP is currently paving the way for the development of relevant regulations and guidelines as a means of implementation instruments for Article (18).

ESTABLISHING AMBIENT AIR MONITORING AND ASSESSMENT NETWORK

The purpose of the Ambient Air Monitoring & Assessment Network is: to judge compliance with and/or progress made towards meeting ambient air quality standards (JS 1140/1996); to observe pollution trends throughout the Kingdom, including non-urban areas; to provide a database for research and evaluation of effects on urban management and landuse and on transportation planning; to develop and evaluate abatement strategies; and to develop and validate diffusion models.

The Network should be designed to meet four basic monitoring objectives: determine highest concentrations expected to occur in the area covered by the network; determine representative concentrations in areas of high population density; determine the impact on ambient pollution levels of significant sources or source categories; and determine general background concentration levels.

Basis for Action

There is a lack of coordination and integration among the institutions involved in air quality monitoring and assessment in the Kingdom. Monitoring of air quality is carried out occasionally and contractually in certain parts of the Kingdom; thus lacking continuity and comprehensive coverage of pertinent areas around the country.

Steady rapid population growth and the consequent increase in the number of vehicles and energy consumption in the Kingdom have led to severe deterioration of air quality in many areas / locations around the country. Rapid industrialization, which led to the overlap of industrial and craft areas with residential and commercial areas, has also contributed to poor air quality in those areas.

NATIONAL CONTROL AND ABATEMENT MEASURES

Basis for Action

Several air pollution sources including mobile and/or stationary ones contribute substantially to the deterioration of air quality in many areas of the Kingdom. Energy generation in the thermal power plant, oil refinery activities and Khirbet Al-Samra wastewater treatment in the Hashimyeh/Zarqa area contribute to elevated levels of sulfur dioxide (SO_2), hydrogen sulfide (H_2S), nitrogen oxides (NOx), carbon monoxide (CO), hydrocarbons (HC), and carbon dioxide (CO_2), all of which adversely affect the inhabitants of these areas. Transportation, as a major source of air pollution in Amman and other major cities, contributes to nitrogen oxides (NOx), carbon monoxide (CO), lead (Pb), sulfur dioxide (SO_2), hydrocarbons (HC), ozone (O_3) and carbon dioxide (CO_2) levels. Several industries contribute to elevating dust levels in the country; these may include phosphate mining, quarrying activities and cement. Other industries such as iron & steel, chemical, and tannery contribute to odor problems and other gaseous pollutants. Wastewater treatment plants and solid waste disposal sites have a major impact on the air quality via emitting odor, CH_4, and H_2S.

PREPAREDNESS AND EMERGENCY PLANS

Basis for Action

There is a lack of comprehensive emergency planning and preparedness on the national level. There is poor understanding of the kind of hazards that might occur with regards to sudden releases of hazardous substances (pollutants), and the consequent risk they may pose for local communities.

GLOBAL COMMITMENTS IN THE FIELD OF ATMOSPHERIC PROTECTION

Basis for Action

Jordan is a signatory of the Vienna Convention, Montreal Protocol, Copenhagen and London Amendments regarding Ozone layer protection. Jordan's consumption of Ozone Depleting substances (ODs) is estimated at 1000 tones/year; thus its consumption is less than 0.3 kg/yr./capita.

MANAGEMENT OF COASTAL ZONES

Only access to the sea; its shoreline is a stretch of about 26 km. The city has witnessed progressive development of its port to a busy city of more than 69,000 inhabitants (1998). The development of industry, tourism and transport has brought about economic advantages, but also has adverse impact on the environment, natural resources and the population; The environmental characteristics and resources of Aqaba are influenced by the development activities in the economic sectors, and by the pollution that follows. The Aqaba coastline contains a very highly diverse marine environment, including the coral reefs. Being the only sea outlet for Jordan, the port of Aqaba has witnessed development of diverse activities; industry and tourism competed to develop the coastline, while port facilities grew to meet the export/ import demands. The Aqaba Regional Authority has developed its own action plan on the environment.

COASTAL ZONE MANAGEMENT AND MONITORING

Basis for Action

The Gulf of Aqaba has a unique biological diversity, which makes it a global priority for preservation. The Gulf is semi-enclosed, a factor that contributes to its rich and unique marine ecosystem, but also results in limited water exchange with the Red Sea. Therefore, it has a low potential for the dispersion of oil and other contaminants. This aspect of confinement and concentration of pollutants is particularly acute in the Jordanian Portion of the Gulf.

The port city of Aqaba and the unique marine environment in the Gulf is of crucial impor'ance to the national interests. However, the development of Aqaba over the years has brought threats to the marine environment such as the construction activities along the shoreline, the phosphate dust during transport and loading/ unloading, ship movement, and population growth. The Aqaba Regional Authority has taken steps towards alleviating some of their problems.

However, continued and systematic efforts are needed for the control of the marine environment. Some issues are to be resolved: Lack of coordination and integration among the institutions involved in marine environment quality monitoring and assessment in the country, which led to the fragmentation in the management of the Coast. The continuous rise in harbor activities, and consequent increase in the deterioration of marine environment quality. The rapid industrialization which led to the overlap of industrial and craft areas with residential and commercial areas also contributed to poor quality to the marine environment in the same areas and imposed difficulties in the management and control.

CONTROL OF URBANRELATED POLLUTION ACTIVITIES

Basis for Action

Aqaba has developed from a small town to a busy city of about 69,000 inhabitants. The city is expected to continue expanding, as more developments occur in the industrial, tourist and transport sectors. The city planners have prepared programs for rehabilitation of the old quarters and surrounding slum areas. However, all development activities are influencing the natural resources and the environment, and may adversely effect the public health. Key issues to be considered: Urban infrastructure requires continued upgrading and expansion. The transport system (especially truck movement) causes road hazards, air pollution and leaves unwanted tires as waste. Tourist expansion has negative impacts on the coastal environment.

PREPAREDNESS AND EMERGENCY PLANS

Basis for Action

The hazards and dangers of accidents are many, and a state of emergency might arise. At the moment there is lack of comprehensive emergency planning and preparedness. In addition, there is a poor understanding of the kinds of hazards that might exist with regards to sudden release of hazardous substances (e.g., oil spills), and the consequent risks that may be imposed on the environment and for the inhabitants.

THE DEAD SEA

The Dead Sea is characterized by a number of factors, some of which bear a potential for development, while others are constraints on any development. The Dead Sea is highly saline (280 gm/kg as has a unique topography and geology, and at Sea Level of (- 410 m /Nov 1996) is the lowest place on earth. However, there are resource constraints that can affect the progress of development; it has a fragile environment that may not sustain the ambition and plans of developers. The mineral activities at the southern part of the sea provide economic benefits to the country; the mineral content of the seawater and mud has a potential as resource material and as a therapèutic value. The potential for tourism is there: the sea itself, the rugged terrain and the archaeological sites. However, the Eastern Shore of the sea remains relatively poor and underdeveloped. The situation is now changing. The Jordan Valley Authority (JVA) has already a master plan for the area with flexibility within the Suweimeh and Zara areas to accommodate proposals by the private sector.

The master plan forecasts the construction of 15,000 rooms and 18,000 housing units along the 60 kilometres beach stretch by the year 2010. Moreover, an extension project is being considered for the Suwimeh area that will cover about (1600) dunums and provide approximately 3305 bed units distributed over hotels (400 bed units), pensions (685 bed units), chalets (1000 bed units), tourist villages (600 bed units), youth campsite (620 bed units). Additionally, there will be general service and recreational facilities, public parks, car parking, commercial centres, public beaches and natural health centres. The following programme areas are proposed for consideration, listed with objectives and activities, to allow for and guide the development process to proceed on a sustainable basis: Little development has occurred on the Eastern Shore of the Sea. In the more recent years, as the prospects of ending the conflicts and establishing peace have improved, new opportunities for development are now emerging. The Dead Sea is already under stress; reduced inflow from the Jordan River over the years took its toll from its surface area and its water.

MANAGEMENT OF INFRASTRUCTURE

Basis for Action

The area bordering the North Eastern Shore of the Dead Sea, comprising three small settlements of about 3,250 population, is underdeveloped, poor and has limited resources. The area also contains a number of archaeological sites, some are partially uncovered, and some still need excavations. The area is marked for development for the eastern shore of the Sea has been proposed earlier by JVA; the development plan calls for a complex of tourist facilities, ranging from hotels, chalets to shopping, recreational and service centres.

An extension of the development plans has been recently reviewed and assessed (EIA prepared in February 1998). The development requires provision of the basic services for the new community. Therefore, a lot of activities are going to take place, during the construction and the operation phases. A number of issues should be considered in the planning and operation of the tourist complex: The area has limited water resources as well as other natural resources. Provision of water and sanitation should cover the local population.

The area has many archaeological sites of varying importance, which will be negatively impacted as a result of the development of the projects in its various stages of construction and operation. The local inhabitants are poor; the socio-economic development of the local communities is essential for the sustainability of the resources in the area and the development plans.

PROTECTION AND CONSERVATION OF THE NATURAL AND CULTURAL HERITAGE

Basis for Action

The North-Eastern area shore of the Dead Sea has a long history of settlement. There are many known sites of archaeological significance, and it is very likely that many other buried sites exist throughout the area. Some sites are already disturbed, whether by looting, illicit digging or encroachment and leveling for agriculture. The proposed development plans for the area will most probably damage the sites, if due consideration is not given in the planning and construction phases.

The eastern coastline of the Dead Sea has witnessed few development activities A coastal road had been built few years agoThe area is characterized by a natural beauty, with a number of wadis opening into the area. Therefore, actions are required for preserving and enhancing the physical and the aesthetic elements of the natural heritage.

PUBLIC HEALTH

In spite of the pressures from increased population growth, the Ministry of Health has succeeded in providing health services to the whole population. Primary health care centres are distributed geographically to meet the needs and to improve the well being of the urban and rural dwellers. Of these 1161 health care centres, 273 are located in the Amman area. Thus, 50% of the population located in the Amman area are served with only 23% of the primary health care centres. This leads to overloading of these centres and to decreased efficiency of service. Both efficiency of those centres, as well as, the qualities of the care they provide needs improvement.

It is thought that insufficient qualification of the paramedical personnel and sometimesunsuitable premises are the reasons behind the poor care provided. In 1997 there were 75 hospitals in Jordan with a capacity of 8129 beds. Out of these hospitals, the Ministry of Health owns 22 with a capacity of 3207 beds (39.5%), 43 hospitals are privately owned with a capacity of 2629 beds (32%). The rest are owned by the military health services (9) hospitals with 1731 beds (22%) and one hospital owned by the University of Jordan with 506 beds (6.2%).

The number of people per hospital bed is 555. Scarcity of fresh water resources is the major constraint to sustainable development in Jordan and to environmental health promotion. However, Jordan has been able to control most of its infectious diseases and has reached very good health indicators compared to other developing countries. Rapid industrialization is now bringing new health problems associated with increased pollution and exposure to toxic chemicals.

This is evident when examining the major causes of death, which are cardio-vascular diseases, respiratory diseases and neoplasm. This shows that the epidemiological profile trends are moving from that of a developing country (predominantly infectious diseases) towards an industrial country profile (chronic diseases as predominant causes of mortality/morbidity: cardiovascular, cancers respiratory deficiencies, mental disorders and accidents).

MEETING THE PRIMARY HEALTH CARE NEEDS PARTICULARLY IN THE RURAL AREAS

Basis for Action

During the last decade, considerable progress was achieved in the health sector in view of the overall strategy which was adopted and implemented by the Ministry of Health (MOH). This progress was an outcome and result of the intensive programs, which were implemented especially in the field of Primary Health Care (PHC). In 1995, the number of PHC centres was 1084 including 11chest centre, 166 dental centres and 287 MCH centres. Progress in the health sector also covered both secondary and tertiary services.

The number of hospitals increased by 8 hospitals in two years (1994, 1995), thus increasing the total bed capacity to 7440 beds in all sectors. The private sector is playing a very important role in the provision of the tertiary treatment levels. Out of the new 8 hospital, 5 were built by the private sector. Despite of these achievements and the restless efforts by MOH, some crucial issues remain to be resolved: Inadequate geographical distribution of the health services especially the PHC services, which are necessary to sustain and promote public health. Out of the 1084 PHC centres, 249 of them are located in the Amman- Zarqa area, which accommodates around 50% of the total population of the country.

This means that 50% of the people depend on only 23% of the services affecting thereby the quality of the services rendered to the public. Decreased efficiency of the health services. It is believed that efficiency of the health centres in providing the necessary services needs to be improved. Decreased efficiency is attributed to both the lack of highly qualified personnel and to the inappropriate premises of some centres.

Mothers and children suffer from dietary deficiencies such as iodine and iron deficiencies. The existing Health Information System in MOH needs support for upgrading and expansion. Accurate and reliable health data are imperative for health assessment and development. The problem of poor and insufficient information is not bound to health, rather, it encompasses several other fields. The discrepancy between the reported number of deaths and the estimated (more realistic figure) numbers is an important example. Information management in all sectors needs to be improved

CONTROL OF COMMUNICABLE DISEASES

Basis for Action

High population growth rates and demographic imbalances have overloaded various services such as safe drinking-water supply and sanitation. One per cent of the population lack basic sanitation provisions and 94% of the treated domestic sewage is reused for irrigation. Environmentally related infectious diseases are a major health issue, especially in rural areas and slums. They are affected by development and depend largely on the availability of safe drinking water and proper sanitation and waste disposal practices.

In addition, the control of disease vectors and immunization programs are very important factors. About 15.2% of the rural population obtain untreated drinking water directly from wells, springs and other unsafe sources. Overcrowding and indoor air pollution encourages the spread of acute respiratory infections, which is a major cause of death of children under five years of age. Control programs started in 1992, and they exert a big burden on the health services. The average occupancy rate in the refugee camps and slum areas is 4 people/room and 3.2 people/room, respectively.

PROTECTION OF VULNERABLE GROUPS

Basis for Action

About 42.7% (1994 estimates) of the population are infants and children under the age of 14. The infant mortality rate has decreased from 151/1000 births for the year 1961 to 28/1000 births for the year 1994. This can be attributed to the improvement of health services and education of mothers. Nevertheless, acute respiratory infections, birth trauma, birth asphyxia and communicable diseases remain the main cause of children mortality. The health of children is vulnerable to the various environmental factors.

The rapid urbanization, which leads to socio-economic changes, makes the youth sector a vulnerable group (13.3% of the population are between the age of 14 and 20) to all the problems associated with the encountered changes. About 48.5% of the population are females and 23% are females between the ages of 15 and 49. The fertility rate has decreased to reach 5.6 births for the year 1991 and it is expected to decrease further to reach 4.01 by the year 2005. This can be attributed to the adequate educational opportunities to women in general.

Illiteracy rate among females is 20.6%. About 15% of the female population participated in the work force for the year 1989. Forced migrations and rural migrations have created 13 refugee camps and 24 slum areas in addition to rural and pre-urban areas. These areas generally lack proper housing, health services, sanitation and safe drinkingwater.

MEETING THE URBAN HEALTH CHALLENGES

Basis for Action

The vast urban growth is attributed to the increased rural migration and the forced migrations in the aftermath of the 1948 and 1967 wars and the 1991 Gulf war. The demographic imbalances have overloaded the capacity to meet human needs and strained services and exposed the population to environmental hazards. Environmental pollution, in addition to overcrowding and inadequate housing contributes to the deterioration of the quality of human health.

REDUCING HEALTH RISKS FROM ENVIRONMENTAL POLLUTION AND HAZARDS

Basis for Action

Rapid industrialization and urbanization took place during the last decade, meanwhile, health and environmental impact assessment was absent. Pollution-related health problems are modest, yet pollution is a key issue to be tackled in development strategies, especially the management of hazardous and chemical wastes. Pollution control, prevention and protection measures have not kept pace with the rapid industrialization and urbanization.

Though some improvement has been achieved, the ability to deal with environmental pollution problems is still limited due to lack of resources, information, best available technologies, follow-up, etc. The increase in the number of motor vehicles and transportation emissions and the absence of air pollution abatement measures at industries, in addition to open burning of domestic wastes, are major causes of outdoor air pollution.

Respiratory diseases and other health problems among workers can be attributed to unsafe conditions and unhealthy indoor air quality at the workplace. Dumping of some industrial wastes and toxic chemicals in an uncontrolled fashion in remote areas, and in sewer manholes, etc. threatens the quality of the environment and the public. Reuse of treated domestic and industrial wastewater, which, if not controlled and monitored properly, can lead to health problems. Improper disposal of hospital waste that can cause major health hazards to exposed groups of the community. Poisoning and chemical accidents due to exposure and misuse of toxic chemicals (e.g. pesticides).

Lack of awareness of the potential health risks and environmental hazards associated with the use of chemicals in the industrial, agricultural, commercial as well as domestic sector. Over pumping of underground water is affecting the quality of water. Industrial and agricultural pollutants can lead to contaminated food, in addition to the possible use of unsafe food additives and colours.

Chapter 5

Phytoremediation

Phytoremediation uses plants to clean up pollution in the environment. Plants can help clean up many kinds of pollution including metals, pesticides, explosives, and oil. The plants also help prevent wind, rain, and groundwater from carrying pollution away from sites to other areas. Phytoremediation works best at sites with low to medium amounts of pollution. Plants remove harmful chemicals from the ground when their roots take in water and nutrients from polluted soil, streams, and groundwater. Plants can clean up chemicals as deep as their roots can grow. Tree roots grow deeper than smaller plants, so they are used to reach pollution deeper in the ground. Once inside the plant, chemicals can be:

- Stored in the roots, stems, or leaves
- Changed into less harmful chemicals within the plant
- Changed into gases that are released into the air as the plant transpires (breathes).

Phytoremediation can occur even if the chemicals are not taken into the plant by the roots. For example, chemicals can stick or sorb to plant roots. Or they can be changed into less harmful chemicals by bugs or microbes that live near plant roots. The plants are allowed to grow and take in or sorb chemicals. Afterwards, they are harvested and destroyed, or recycled if metals stored in the plants can be reused. Usually, trees are left to grow and are not harvested. Plants grown for phytoremediation also can help keep harmful chemicals from moving from a polluted site to other areas.

The plants limit the amount of chemicals that can be carried away by the wind or by rain that soaks into the soil or flows off the site. Before phytoremediation begins, EPA studies whether plants grown to clean up pollution can be harmful to people. EPA tests the plants and air to make sure that the plants do not release harmful gases into the air. Some insects and small animals may eat the plants used for phytoremediation. Scientists are studying these animals to see whether the plants can harm them. Scientists also are studying whether these animals pose harm to the larger animals that eat them. In general, as long as plants are not eaten, they are not harmful to people.

The time it takes to clean up a site using phytoremediation depends on several factors:

- Type and number of plants being used
- Type and amounts of harmful chemicals present
- Size and depth of the polluted area
- Type of soil and conditions present

These factors vary from site to site. Plants may have to be replaced if they are destroyed by bad weather or animals. This adds time to the cleanup. Often it takes many years to clean up a site with phytoremediation. EPA uses phytoremediation because it takes advantage of natural plant processes. It requires less equipment and labour than other methods since plants do most of the work.

Trees and plants can make a site more attractive as well. The site can be cleaned up without removing polluted soil or pumping polluted groundwater. This allows workers to avoid contact with harmful chemicals. Phytoremediation has been successfully tested in many locations, and is being used at several Superfund sites.

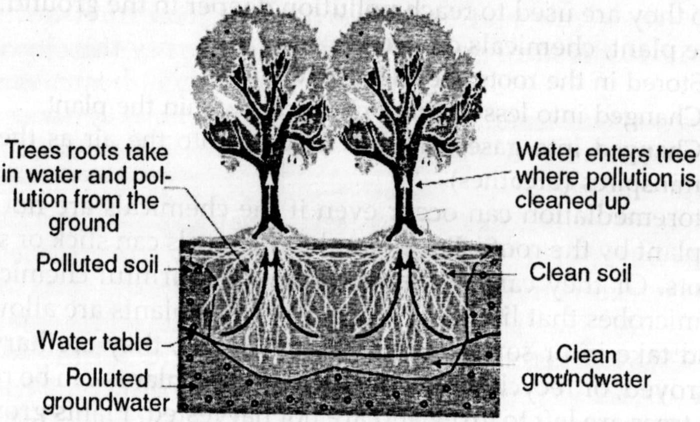

PHYTOREMEDIATION OF TCE USING POPULUS OVERVIEW OF PHYTOREMEDIATION

Phytoremediation is an emerging technology which uses plants and their associated rhizospheric microorganisms to remove, degrade, or contain chemical contaminants located in the soil, sediments, groundwater, surface water, and even the atmosphere. Researchers have found that plants can be used to treat most classes of contaminants, including petroleum hydrocarbons, chlorinated solvents, pesticides, metals, radionuclides, explosives, and excess nutrients. Plant species are selected for phytoremediation based on their potential to evapotranspirate groundwater, the degradative enzymes they produce, their growth rates and yield, the depth of their root zone, and their ability to bioaccumulate contaminants.

Table lists the various applications of phytoremediation technologies. This list indicates that phytoremediation is actually a broad class of remediation techniques which include many treatment strategies. Obviously, the common thread through all of these techniques is the use of plants to treat a contaminant problem. However, due to the diverse nature of chemical contamination and the diversity of plants with the potential to treat them, remedial project managers must choose between a wide variety of phytoremediation techniques to solve the problem at hand.

Despite the diversity of phytoremediation technologies, its application is limited by a number of factors. Phytoremediation can only work at sites that are well suited for plant growth. This means that the concentration of pollutants cannot be toxic to the plants, and the pollution cannot be so deep in the soils or groundwater that plant roots cannot reach it. As a result, phytoremediation may be a good strategy for sites conducive to plant growth with shallow contamination, it may be a good secondary or tertiary phase in a treatment train for highly polluted sites, or it may not be a viable option for a site.

Table. Type of Phytoremediation System

Treatment Method	Mechanism	Media
Rhizofiltration	Uptake of metals in plant roots troughs	surface water and waterpumped through
Phytotransfor-mation	Plant uptake and degrad-ation of organics	surface water, ground-water
Plant-Assisted Bioremediation	Enhanced microbial degra-dation in rhizosphere	soils, groundwater, within the rhizosphere
Phytoextraction	Uptake and concentration of metals via direct uptake into plant tissue with sub-sequent removal of the plant	soils
Phytovolatilization	Root exudates cause metals to precipitate and become less bioavailable	soils, groundwater, mine tailings

Phytovolatilization	Plant evapotranspirates mercury, and volatile organics	soils, groundwater
Removal of organics from the air	Leaves take up volatile organics	air
Vegetative Caps	Rainwater is evapotransp-irated by plant to prevent leaching contaminants from disposal sites	soils

Even though phytoremediation appears to have limited application, researchers in industry, academia, and government are looking into phytoremediation as a useful treatment technology. In fact, a number of companies that offer phytoremediation technologies have been started in the last few years, and many larger consulting firms are beginning to offer phytoremediation services as well.

Again, since phytoremediation covers a broad spectrum of pollutants and treatments, many of these companies focus most of their attention on one niche of the phytoremediation field (e.g., metal extraction from soils, poplar tree buffers, etc.). Most of the phytoremediation companies have already used, or are using, phytoremediation in the field.

For example, four phytoremediation projects have been accepted for the Superfund Innovative Technology Evaluation (SITE) programme. One project by Phytotech, Inc. involves the use of plants to extract metals from soils. The second project by Phytokinetics involves the use of poplar trees to treat PAH contaminated groundwater. Another project by Phytokinetics uses grasses to remediate surficial soils contaminated with PCP and PAHs.

The fourth phytoremediation SITE project involves the use of cottonwoods to treat a plume of TCE in shallow groundwater at the Carswell Air Force base in Ft. Worth, TX. This last project will be detailed as a case study later in this report. In addition, a number of other pilot and field scale projects have taken place around the country. Appendix B summarizes some of these projects, including the types of contaminants present and treatment techniques employed at these sites.

ADVANTAGES AND DISADVANTAGES OF PHYTOREMEDIATION

Early research indicates that phytoremediation technology is a promising cleanup solution for a wide variety of pollutants and sites, but it has its limitations. Table summarizes some of the advantages and limitations of phytoremediation. An examination of the table reveals that many of phytoremediation's limitations and advantages are a direct result of the biological aspect of this type of treatment system. Plant-based remediation systems can function with minimal maintenance once they are established, but they are not always the best solution to a contamination problem.

One way to summarize many of phytoremediation's limitations is that the pollutant must be bioavailable to a plant and its root system. If a pollutant is located in a deep aquifer, then plant roots cannot reach it. If a soil pollutant is tightly bound to the organic portion of a soil, then it may not be available to plants or to microorganisms in the rhizosphere. On the other hand, if a pollutant is too water soluble it will pass by the root system without any uptake.

Table. Advantages and Limitations of Phytoremediation

Advantages of Phytoremediation	Limitations of Phytoremediation
In situ	Limited to shallow soils, streams and groundwater
Passive	High concentrations of hazardous materials can be toxic to plants
Solar driven	Mass transfer limitations associated with other biotreatments
Costs 10% to 20% of mechanical	Slower than mechanical treatments treatments
Transfer is faster than natural attenuation	Only effective for moderately hydrophobic contaminants
High public acceptance	Toxicity and bioavailability of degradation products is not known
Fewer air and water emissions	Contaminants may be mobilized into the groundwater
Generate less secondary wastes	Potential for contaminants to enter food chain through animal consumption

Performance

A major hurdle for innovative technologies to overcome is a lack of performance data, and phytoremediation is no exception. One of the current barriers to performance data is the length of time involved in a phytoremediation project. Plants can only grow so fast, so obtaining long term performance results is dependent on the rates of plant growth and activity. There are currently a number of pilot scale projects in existence, but they have not resulted in conclusive performance data at this time. These sites are being monitored and will report results over the next few years. Also, a number of firms have installed phytoremediation systems at polluted sites owned by private clients, so results from those sites are not publicly available. Some performance data is included in the "Case Studies" section of this report. However, these studies are not complete, so the data presented is from early in those treatments.

When looking at performance data, one needs to keep in mind that most data is very site specific, especially for a biological system like phytoremediation. For example, transpiration rates reported for a hot, arid region may not match those from a cooler climate. Performance may also be seasonal since many plants are dormant during the winter months. This is especially important when planning projects or comparing results of projects from different regions.

COST

In addition to performance data, accurate cost data is often difficult to predict for new technologies. Most lab, pilot, and field scale tests include monitoring procedures far above those expected at a site with a remediation goal. This inflates the costs of monitoring at these test sites. As a result, it is difficult to predict the exact cost of a technology that has not been established through years of use.

However, since phytoremediation involves the planting of trees or grasses, then it is by nature a relatively inexpensive technology when compared to technologies that involve the use of large scale, energy consuming equipment. Phytoremediation costs will vary depending on the treatment strategy. For example, harvesting plants that bioaccumulate metals can drive up the cost of treatment when compared to treatments that do not require harvesting. Regardless, phytoremediation is often predicted to be cheaper than comparable technologies.

Table presents some estimates of phytoremediation's costs in relation to conventional technologies. This table represents some vague and variable estimates due to the current dearth of cost information. Since the bulk of this report deals with the use of poplar tree systems, Table lists some of the costs listed by two companies who specialize in poplar designs.

Keep in mind that costs of phytoremediation are highly site specific, so that any estimate found in these tables is merely a rough estimate of potential costs. Many of these estimates are speculative based on laboratory or pilot scale data.

Table. Estimates of Phytoremediation Costs Versus Costs of Established Technologies

Contaminant	Phytoremediation Costs	Estimated Cost using Other Technologies	Source
Metals	$80 per cubic yard	$250 per cubic yard	Black (1995)
Site contaminated with petroleum hydrocarbons (site size not disclosed)	$70,000	$850,000	Jipson (1996)
10 acres lead contaminated land	$500,000	$12 million	Plummer (1997)
Radionuclides in surface water	$2 to. $6 per thous-nd gallons treated	none listed	Richman (1997)
1 hectare to a 15 cm depth (varopis (contaminants)	$2,500, $15,000	non listed	Cunningham *et al.* (1996)

Phytoremediation of TCE with Trees in the Genus Populus

Researchers have been investigating the possibility of using trees in the genus Populus to hydraulically contain and ultimately remediate plumes of TCE in groundwater. In addition, managers at several Department of Defence and Superfund hazardous waste sites have planted Populus sp. in an effort to treat groundwater contaminated with chlorinated solvents.

This section will focus on the available information about the use of Populus sp. to phytoremediate plumes of TCE. Included are background sections on TCE and poplar trees, followed by sections on mechanisms of TCE phytoremediation, other phytoremediation applications of poplars, and a comparison of conventional remediation methods to poplar systems.

Description of TCE

Trichloroethylene (TCE) is a common contaminant at many of the nation's hazardous waste sites. It can be found at 50% of Superfund National Priority List (NPL) sites with completed Records of Decision (RODs), and it is above action levels in the groundwater of 17% and soils of 16% of RCRA corrective action facilities. TCE pollution became prevalent primarily through its use as an industrial degreasing agent.

Other uses of TCE include its use as a solvent for dry cleaning, an anaesthetic for medical and dental use and as an ingredient in paints, inks, cosmetics, disinfectants, and cleaning fluids. Due to widespread TCE contamination, finding innovative ways to clean this pollutant has become a priority in the remediation field.

Physical and Chemical Properties of TCE

Examining the physical and chemical characteristics of a target compound is very important when choosing a remediation strategy. The physical characteristics of TCE, which are listed in Table, make it difficult to remove from the groundwater using traditional technologies. TCE is a dense non-aqueous-phase liquid (DNAPL), meaning that it is denser than water and therefore tends to exist in undissolved pools in the bottom an aquifer.

This property makes it very difficult to treat TCE by methods such as pump and treat because it is almost impossible to tap into small pools of undissolved TCE that reside in the groundwater. Pump and treat methods can remove TCE that is in the aqueous phase, but since pools of non-aqueous TCE are in equilibrium with the groundwater, more TCE will dissolve into the aqueous phase as the groundwater is treated. This results in a continuous cycle of slow dissolution from the nonaqueous to the aqueous phase that could take many years and large amounts of money and energy to treat.

Table. Physical and Chemical Properties of TCE

Property	Value
Molecular weight	131.5
Boiling point	87°C
Melting point	– 73°C
Specific gravity	1.4642 at 20°C
Solubility in water	1,000 mg/litre
Log octanol/water partitioning coefficient	2.29
Vapour pressure	60 mm Hg at 20°C
Vapour density	4.53

Many scientists and hazardous waste site managers, such as those involved with the three case studies included in this report, believe that it is more efficient to use trees as a solar driven pump and treat mechanism in a long term treatment process for TCE contamination. Others, however, remain skeptical. Cunningham *et al*. stated that TCE and perchloroethylene (PCE) are "a relatively poor choice of targets for phytoremediation" because they tend to form dense pools near the bottom of an aquifer, out of the reach of tree roots. Current studies will likely provide some answers to this debate.

TCE Availability to Plant Roots

In order for a plant to directly degrade, mineralize, or volatilize a compound, it must be able to take that compound up through its roots. The ability of a plant to take up a chemical from the soil and groundwater and translocate it to its shoots is described by a chemical's root concentration factor (RCF) and transpiration stream concentration factor (TSCF). The RCF is a measure of the root concentration of a contaminant versus the concentration in the external solution, while the TSCF is a measure of the concentration in the xylem sap in relation to the concentration in the external solution. Both of these factors vary directly with a chemical's water solubility, commonly expressed as its log K_{ow}. According to Briggs *et al.*, contaminants in solution with the highest TSCF are moderately soluble compounds with a solubility in the range of log K_{ow} 1.5 to 2.

However, several reviews expand this optimum range to include log K_{ow} values as low as 0.5 and as high as 3. Most chlorinated solvents, including TCE, fall within this expanded range, along with BTEX chemicals and short chain aliphatics. As a result, phytoremediation appears to be a viable option for treating dissolved TCE in groundwater. Soils, on the other hand, pose a potential problem for plants because may they contain high levels of organic matter.

In soils, the log K for maximum TSCF can be shifted down to favour more polar molecules because of the competing process of sorption to soil organic matter. As a result, plant roots alone may have a difficult time extracting TCE from soils containing significant amounts of organic matter. However, microorganisms in the rhizosphere are capable of degrading TCE. Anderson and Walton found that TCE mineralization was significantly enhanced in the rhizoshpere of a number of plant species, such as Pinus taeda (loblolly pine) and Lespedeza cuneata, when compared with nonvegetated soils. The exact mechanisms of rhizospheric degradation will be discussed in a later section

Toxicity of TCE to Animals

TCE has been found to be carcinogenic to laboratory mice in a number of studies reported by the National Institute of Environmental Health Sciences (T-2, TR-243). However, these studies did not find TCE carcinogenic to laboratory rats. TCE was found to be a mutagen when tested using microbial assay systems, and chronic inhalation exposure to TCE causes liver, kidney, neural, and dermatological reactions in animals. The EPA drinking water standard for TCE has been set at 5 parts per billion (ppb), and the EPA has also reported that drinking 1 part per million (ppm) TCE in water over a lifetime will cause 32 humans in a population of 100,000 to be at risk of cancer.

Vinyl chloride, which may result from the anaerobic breakdown of TCE by microorganisms, is regarded as a more potent human carcinogen than TCE. The EPA drinking water standard for vinyl chloride is 2 ppb, and the EPA estimates that drinking 1 ppm vinyl chloride over a lifetime will cause 9,570 cases of cancer in a population of 100,000 people. Since this microbial breakdown product is more chronically toxic than its parent compound, regulators are concerned with the exact fates of TCE in a remediation system.

Toxicity of TCE to Plants

While it is important to understand the concentrations of TCE that are toxic to plants, few studies report the phytotoxic effects of TCE. Gordon *et al.* reported that poplars were able to survive when grown in water containing 50 ppm TCE. Another experiment found TCE to be acutely toxic to a variety of crop plants at concentrations of about 2 mM in the gas phase.

The later study hypothesized that an increase in electrolyte leakage or interference with the photosynthetic system was the mechanism of acute toxicity in the plants, but the exact mechanism is not known.

Description of Populus

The genus Populus includes a number of species of trees such as poplars, cottonwoods, and aspens. Populus is a member of the Salicaceae family, which also includes willows. There are around 30 species of Populus distributed around the Northern Hemisphere, with eight species indigenous to North America and others that have been introduced. In addition, Populus sp. have the ability to cross within the genus both in the wild and through controlled breeding, so there are a large number of potential hybrids.

Due to their ability to readily form hybrids, poplars have been crossed by foresters for years in order to maximize growth rates and yield. Hybrid poplars were originally bred and grown as a cash crop for such uses as pulp wood and as a renewable energy source, but because of their rapid growth rates and high evapotranspiration rates, they make ideal candidates for phytoremediation. Table summarizes many of the advantages of using poplars for phytoremediation.

Advantages of Populus sp. in Phytoremediation:

- Greater than 25 species worldwide
- Fast growing (3 to 5 meters/year)
- High transpiration rates (100 liters/day optimally for 5 year old tree)
- Not part of food chain
- Trees can be used for paper production or as biomass for energy
- Longlived (25-30 years)
- Grow easily from cuttings
- Can be harvested and than regrown from the stump

The goal of poplar hybridization is to achieve heterosis, which means that the genetic traits of hybrids exceed those of the parents. Two species of poplars, Populus deltoides (eastern cottonwood) and Populus trichocarpa (black cottonwood), are commonly crossed for use in phytoremediation. Populus trichocarpa x deltoides, have leaves that are about four times as large as the leaves of parent plants (ORNL 1996). Increasing leaf size increases the potential evapotranspiration rates of these trees due to increased total leaf surface area.

Another common cross that has been used in phytoremediation studies is P. deltoides x P. nigra (black poplar). This cross is sometimes referred to as P. x euramericana due to the original distribution of the two species, the black poplar in Europe and the eastern cottonwood in North America. There are many other hybrids of poplars that have been developed, some of which have been or will likely be used in phytoremediation systems. One piece of information that concerns most of those interested in poplar phytoremediation projects is the evapotranspiration rates of the trees.

Poplars and cottonwoods are phreatophytic plants, which means that they can extend their roots to the water table and pump from the zone of saturation. For this reason, the presence of a number of cottonwood species in desert regions, such as Populus fremontii and Populus wizlizeni, has been historically used as an indicator of relatively shallow groundwater. This ability to pump groundwater has earned poplars the name "solar driven pump and treat systems" in the phytoremediation field. Knowing how well these trees can act as a solar powered pump will aid in deciding whether or not they can be used to treat a particular site.

A plot of trees with high evapotranspiration rates can cause a significant draw-down in the water table, which results in a hydraulic barrier to contaminant transport. Results of a number of studies indicate that a stand of poplars can cause a depression in the water table ranging from several inches to several feet.

Unfortunately, since trees are solar driven and biological, as opposed to mechanical, pumping rates will vary with tree age and species as well as time of day, time of year, amounts of solar radiation, and other climatic and geographic factors. As a result, pumping rates will be highly site specific. Table lists some reported evapotranspiration rates for individual poplar trees. This list includes a wide range of pumping rates, from 1.6 gpd/tree for a young tree to 53 gpd/tree for a five year old poplar.

Table. Estimates of Evapotranspiration Rates by Hybrid Poplars

Rate	Source
100 to 200 L/day/tree (~26 to 53 gallon/day) for 5 year old trees	Newman *et al* (1997)
100 L/day/tree for a 5 year old tree under optimal conditions	Stomp *et al.* (1994)
13 gallons per day (estimated) when trees are calculated as low-flow pumping wells	Sheldon Nelson-Workshop on Phytoremediation of organic Contaminants (1996)
1.6 to 10 gpd/tree (observed) sap flow rates for young hybrid poplars at the Aberdeen Proving ground sin Maryland	Compton (1997)
10-11 kg/tree/day (observed) in early summer for 1-2 year old Easterncottonwoods growing in Taxas	Gred Harvey (Personal Communication
40 gallons per day (observed) for 5 year old trees in Utah in the summer of Organic	Ari Ferro - Workshop on Phytoremediation Contaminants (1996)

Trees will transpire at different rates when grown together as a stand than they would when grown individually. This is because evapotranspiration varies with the total leaf surface area, whether it be for an individual tree or an entire stand. This means that tree density is an important consideration for phytoremediation. A dense stand will have less leaf surface area per tree, but the combined leaf surface area of a dense stand will be greater than the combined surface area of a thin stand. Some information on evapotranspiration is available for stands of poplar trees. For example, Gordon *et al.* reported that a stand of 5 year old poplars could cause a 140 cm/year draw-down in the water table when grown at a density of 1,750 trees/ha in the warm, arid conditions of eastern Washington state. Table provides measured water uptake per hectare and per acre of poplars grown at a density of 2,170 trees/acre at varying tree ages. This table, along with Table, provides some good approximations of the pumping efficiencies of Populus sp. However, evapotranspiration rates will vary from site to site.

Table. Growth and Water Uptake Potential in Five Growing Seasons in Amana, IA for Poplars Planted at 2,170 Trees/acre Density

Growing Season	Water Uptake (litre/hectare)	Water Uptake (gallons/acre)
1	437,545	46,766
2	2,99,035	319,795
3	8,329,440	890,258
4	9,957,364	1,064,264
5	21,845,073	2,334,847
5 year average	8,712,291	931,188

Mechanisms of TCE Phytoremediation by Populus sp.

The following sections will discuss the phytoremediation mechanisms that have been reported when poplars are used to treat TCE. However, the use of these plants to degrade TCE is still a relatively new idea, so not all of the mechanisms are clearly understood at this time. Therefore, the following sections will outline the current body of knowledge on the subject of TCE remediation mechanisms by Populus sp. This section begins with a brief overview of Populus remediation mechanisms, followed by individual sections detailing what is known about those mechanisms. Also included are some results of controlled field trials of TCE phytoremediation using poplars.

Overview of Mechanisms

A recent study by Newman *et al.* investigated phytoremediation of TCE using two varieties of hybrid poplars (P. trichocarpa x P. deltoides and P. trichocarpa x P. maximowiczii). These experiments were conducted using axenic poplar cell cultures and whole poplar trees grown in a greenhouse. The investigators reported the formation of TCE metabolites in hybrid poplar tree tissues. They confirmed that the trees were responsible for the metabolites in their tissue, and not microorganisms, by finding the same metabolites in sterile poplar cell cultures. They also found that TCE was evapotranspirated by the trees in the greenhouse studies, and that some TCE was incorporated into an insoluble residue within the trees. In addition, Walton and Anderson reported that TCE degradation by microbial organisms was enhanced in the rhizosphere of various plant and tree species.

Table. Mechanism of TCE Phytoremediation by the Populus sp.

Process	Product
Metabolism[2]	chloral hydrate, trichloroethanol di- and trichloroacetic acid
Incorporation[2]	Insoluble residue
Mineralization[2]	CO_2
Transpiration[2]	TCE vapour
Rhizospheric degradation via microorganisms	CO_2[b], dehalgenation metabolites such as *cis*-1,2-dichloroethylene[c], vinyl chloride[d], and others

Enzymatic Degradation and Mineralization in Populus

As stated in the previous section, poplar trees have been found to degrade TCE, but the exact metabolic mechanism or mechanisms of enzymatic degradation is currently under some speculation. Two lines of research have been reported to date. Both are similar in that they indicate an enzymatic process involved in oxidizing TCE to various metabolites. Regardless of the enzyme or enzymes involved, the oxidative process will ultimately mineralize the carbon in TCE to CO.

Also, there is always the possibility of more than one mechanism taking place 2 within a plant or between plant species. The research of Newman *et al.* suggests that TCE metabolism in poplars may be similar to the mammalian breakdown of TCE. This belief is based on the production of similar TCE metabolites in both plants and mammals. However, the exact mechanism was not determined by these researchers, only hypothesized based on the presence of the metabolism products.

According to a review by Cunningham *et al.*, many of the enzyme systems involved in mammalian metabolism of TCE are also found in plants (e.g., cytochrome p- 450 oxygenases and glutathione S-transferases), so this hypothesis seems possible. Another line of research indicates that TCE metabolism in poplars is the result of a dehalogenase enzyme. According to Dr. Laura Carreira, who has isolated the enzyme, dehalogenase is an ethylene degrading enzyme (personal communication). It oxidizes alkanes, alkenes, and methanes and their halogenated analogues.

Dehalogenase will ultimately mineralize TCE to CO via an oxidative pathway. Dr. Carreira has developed an 2 antibody assay for the ethylene degrading enzyme to use as an indicator of its presence in various plant species. This antibody technique can be used to predict the ability of a plant or tree species to degrade chlorinated solvents before that organism is chosen for use at a particular phytoremediation site. In the case of poplars, some species or hybrids can produce more of this enzyme than others, and some may not produce the enzyme at all. As a result, using different hybrids or species at various sites without assaying for the dehalogenase may yield contrasting results.

A tree that manufactures large quantities of the enzyme will have the ability to degrade TCE, while plants that produce little or no quantities of the enzyme will tend to volatilize it. A wide variety of plants can potentially produce the dehalogenase enzyme, so Dr. Carreira's antibody technique may be used to find species of ethylene degrading plants or trees that are native to a site. Table lists some species that have been reported to produce dehalogenase enzyme and the half lives of hexachloroethane, a halogenated hydrocarbon, in the presence of these plants.

Table. Some Plant Species Containing Dehalogenase

Plant Species	Half-life (hours) of Hexachloroethane
Algae Nitella (stonewort)	90
Anthrocerotae sp.	120
Algae Spirogyra	95
Myriophyllium spicatum (parrot feather)	120
Populus sp.	50

Enhanced TCE Degradation and Mineralization in the Rhizosphere

The root zone of plants provides an environment conducive to the growth and activity of microorganisms. These enhanced populations of microorganisms have been found to degrade TCE in the rhizosphere of a number of plant species growing on contaminated sites. In addition to bacteria, microrhizal fungi are capable of metabolizing chlorinated organics.

Initial research determined the plant species capable of promoting TCE degradation in soils. Recent studies have focused on determining the mechanisms of degradation within the rhizosphere. A review on this subject by Davis *et al.* reports that microbial degradation of TCE can take place either aerobically or anaerobically.

The aerobic process is an oxidative mechanism catalyzed by a mono-oxygenase enzyme. Methane mono-oxygenases (MMO) and alkene oxygenases are reported as the primary enzymes involved in the oxidative mechanism. Each enzyme uses either methane or an alkene as its primary substrate, but can also oxidize TCE in a fortuitous reaction. Plants may support this mechanism by transferring exudates to anaerobic sites, stimulating methanogens to produce methane. The methane in turn stimulates aerobic methanotrophs, who cometabolize TCE via the MMO enzyme. Products of this reaction include chloral, dichloroacetic acid, trichloroacetic acid, trichloroethanol, and ultimately CO_2. The other microbial mechanism responsible for TCE metabolism is dechlorination catalyzed anaerobically via a dehalogenase enzyme. The review by Davis *et al* also describes this mechanism in the rhizosphere, which is primarily carried out by methanogens.

Contaminants such as TCE have too low an energy yield to stimulate a population of methanogenic organisms, but plant exudates can supply enough carbon for use as a reductant by the microbes. Once a microbial population increases in density, it will begin to cometabolically dehalogenate TCE. The common products of this reaction are dichloroethylene, vinyl chloride, and eventually ethene and ethane.

Insoluble Residues of TCE

The research of Newman *et al.* also found that a small percentage (3-4%) of TCE remained as an insoluble residue in the poplar tree cell. It is believed that this is due to abiotic binding of TCE to the cell walls of the plant, but the exact mechanism is still under investigation.

TCE Volatilization

Understanding TCE volatilization rates in poplar trees is critical for this technology to gain widespread acceptance amongst hazardous waste site managers and regulators. There may be concern if the trees are transpiring high concentrations of TCE into the atmosphere, where the pollutant becomes an air quality concern. Proponents of phytoremediation argue that VOCs will volatilize from the groundwater, through the soil, and into the air in the absence of trees. For example, plants will invade sites that are left unattended for extended periods of time, and invasive plants may evapotranspirate the contaminant. That being the case, there would be some evapotranspiration in the absence of a treatment strategy.

Phytoremediation schemes would only accelerate the process of volatilization that occurs naturally. Still, volatilization concentrations are decreased by a number of factors, such as exclusion of nonpolar compounds at the roots. According to Davis *et al.*, "Very few contaminants are sufficiently water soluble, nontoxic to plants, and volatile enough to reach atmospheric concentrations that would be of concern by [evapotranspiration]." Despite the fact that evapotranspiration rates are still unclear, Davis *et al.* used energy input estimates to calculate a maximum transfer rate of TCE to the atmosphere. They predicted a maximum transfer rate of 10 g/m^2/day. This estimation assumes that the water is totally saturated with TCE at 1.5 g/L and the TSCF of TCE is 0.67, based on the equations of Briggs *et al.* Using a more realistic groundwater concentration of 1-15 mg/L TCE and a mixing height of 100-300 meters in the atmosphere, the investigators estimated that transfer to the atmosphere would be 4 to 6 orders of magnitude smaller than the maximum. The result is a very low air concentration of TCE downwind in a worst case scenario.

FIELD TRIALS

Investigations were conducted to follow up the work of Newman *et al.* by conducting field trials of TCE phytoremediation in the state of Washington. These field experiments were conducted by Occidental Chemical Corporation along with researchers from the University of Washington and Washington State University. The state of Washington approved a two year field experiment where TCE was added to 3.7 × 6.1 meter cells that were 1½ meters deep and double lined with polyethylene. Hybrid poplars (Populus tricocarpa x deltoides, H-11-11) were planted in the experimental cells and soil was only added to the controls. TCE was added to the cells at a concentration of 50 ppm in the water.

The investigators found that over 95% of the TCE was removed from the stream water in the plots with trees. During the first year of the trial, however, 65% of the added TCE was removed in the control plot without trees. This probably meant that a significant portion of the TCE was bound in the soils (Newman *et al.*1997b).

At the end of the second growing season, 65-70% of the added TCE remained in the water stream in the control cells (Workshop on Phytoremediation of Organic Contaminants), indicating that the loss to the soil decreased substantially after the soil became saturated with TCE. Still, during the second year over 97% of the added TCE was removed from the water stream in the cells containing trees. The investigators also found products of anaerobic microbial dehalogenation, such as three isomers of dichloroethylene and small amounts of vinyl chloride, in the water streams.

Uncertainties of Phytoremediation with Populus sp

Research indicates that phytoremediation of TCE using hybrid poplars will work. The question that remains to be answered is, "To what extent does it work?". Unfortunately, there is no answer to that question at this time. The case studies presented later in this report represent some of the most current data on pilot studies. However, these sites are still very new and phytoremediation is a long, slow process. Currently, managers at these sites do not know if poplars can clean TCE plumes to regulatory standards or how long it will take.

Results of these studies are at least a year or two away, possibly more. Another uncertainty surrounding poplar phytoremediation systems is the fate of the contaminant.

Only a few mass balance studies of TCE fate in the field have been attempted. Mass balance predictions are further complicated by the fact that field conditions will vary greatly between sites and geographic regions.

In addition, different species or hybrids of Populus sp. will have variable abilities to treat TCE in the groundwater. Again, results from the three case studies in this report as well as basic research currently taking place will address some of these uncertainties.

Other Types of Populus Phytoremediation Projects

Populus sp. have been used to treat contaminant problems other than TCE in groundwater. One project in Iowa investigated the ability of poplar strips to act as a buffer to protect water bodies from nutrient runoff. A similar study in Iowa looked at the ability of poplars to buffer triazine pesticide runoff from an agricultural field. Both of these studies indicated that poplars could successfully act as a buffer, although they were less effective at buffering nutrients once the trees dropped their leaves in the fall.

In addition, hybrid poplars are currently being used to act as a hydraulic barrier to contain a plume of gasoline and diesel fuel in the groundwater at a site in Ogden, UT. Another way that poplars have been used is as a component of vegetative caps for landfill facilities. Poplars are also being tested for their ability to phytostabilize metals such as lead, and they have been planted as a part of a constructed wetland design to treat explosives such as TNT and RDX in the soil. There is also a report on the use of poplars to phytoextract zinc from soils.

Populus Phytoremediation Versus Other Treatment Technologies

For those unfamiliar with groundwater treatment options, Table lists many of the other technologies that have been used to treat plumes of chlorinated solvents. Table includes the cost of cleaning a plume of chlorinated solvents using these technologies under an idealized set of conditions. Unfortunately, phytoremediation was not included in this economic analysis.

However, phytoremediation would be somewhat more expensive than natural attenuation because it involves tree planting and maintenance as well as monitoring, and it would be significantly less expensive than pump and treat due to decreased energy needs. Therefore, the costs of a poplar phytoremediation system would likely fall somewhere in the middle of Table, and probably near the less expensive end of the spectrum.

Table: Estimated Costs of Treating PCE in the Groundwater

Treatment Technology	Total Present cost (× $1,000)	Cost/Pound PCE removed	Cost/1,000 Gallons Treated
Pump and treat with air stripping and carbon absorption	$9800	$1600	$8.90
Iron reactive barrier	$3900	$640	$5.30
Biobarrier (substrate enhanced anaerobic bioremediation	$3100	$520	$4.20
In situ bioremediation (substrate enhanced, recirculating sourcezone	$1300	$220	$1.80
Natural attenuation (intrinsic bioremediation)	$890	$150	$1.20

Since poplar phytoremediation systems are primarily used as hydraulic barriers and solar powered pumps, the most closely related engineering technology is a pump and treat system.

The major advantage of poplars over pumps is that poplars provide their own energy, where pumps often consume large amounts of electricity. This could save tremendous amounts of money over the course of a long remediation project.

On the other hand, a major disadvantage of poplars is the fact that their pumping rates vary over the course of a year. In addition, poplar systems will only work at sites where the groundwater contamination is within reach of their roots.

Despite some limitations of poplars when compared to pump and treat, the potential economic advantages of this treatment system are tremendous. A report by the Environmental Security Technology Certification Programme (ESTCP) estimates that poplar trees can be used to treat contaminants such as chlorinated solvents and petroleum hydrocarbons at 1,000 DOD cleanup sites around the world (ESTCP). This could save the government hundreds of millions of clean-up dollars.

CASE STUDIES

Since poplars have been shown to remediate TCE under controlled experimental settings, several Department of Defence and Superfund sites are conducting pilot scale phytoremediation projects. The goals of these projects are to use poplars to remediate plumes of chlorinated solvents in the groundwater. Table provides some general information about each site, including contacts. All three were still in early stages of sampling at the time of this writing, so very little performance data is available.

However, results to date indicate that the phytoremediation systems are working. For example, several of the sites are reporting a depression in the water table beneath the plots and some contaminant volatilization from the trees. Future monitoring will help determine the extent of TCE removal that can be achieved using this technology.

Table. Case Studies Overview of Sites

Site	Size of Planting on Site	Number of Trees Planted	Species or Hybrid
Aberdeen Proving Ground Grounds - J Fields Site	~ 1 acre	183	Populs trichocarpa x deltoides, HP-510
Edward Seas Properties	~ 1/3 acre	118 deep root ~90 shallow rooted	Populus charkowiiensis x incrassata, NE 308
Carswell Air Force Base	~1 acre	660	Populus deltoides

Tables provide some meteorological data for the geographic regions of each site. While climate information will not be directly used in this report, it may be important in the future when people are comparing the effectiveness of treatment strategies at these three sites. The information in the following two tables only provides yearly averages, not data recorded at the sites.

When data from these projects are analysed, actual weather data taken during the treatments will be compared to determine the extent of climate's effect.

Data in Tables and indicates that Edward Sears and Aberdeen, which are both located in the mid-Atlantic region, have similar temperature and rainfall averages. Carswell, which is located in Texas, has a warmer climate and less rainfall.

Table. Average Temperature in °F for Geographic Regions of the Case Study Sites

Site	Jan	Feb	May	Apr	May	Jun	Jul	Aug	Sep	Oct	Nov	Dec	Year
Aberdeen	2.0	3.9	3.7	3.5	2.6	6.2	2.1	0.8	2.7	1.8	5.1	2.9	37.2
Ed Sears	3.7	3.1	4.1	3.9	3.4	3.1	4.1	5.4	3.3	3.2	3.8	3.7	44.9
Carswell	1.8	2.2	2.8	3.5	4.9	3.0	2.3	2.2	3.4	3.5	2.3	1.8	33.7

Table. Average Rainfall in Inches for Geographic Regions of the Case Study Sites

Site	Jan	Feb	May	Apr	May	Jun	Jul	Aug	Sep	Oct	Nov	Dec	Year
Aberdeen	32	35	44	53	64	72	76	75	69	57	48	37	55
Ed Sears	31	33	42	50	60	70	75	74	67	56	47	37	53
Carswell	45	50	56	66	74	82	86	85	78	68	56	48	66

There is some cost information available for each site, but any cost data is skewed because these are pilot scale projects. The projected costs include monitoring and analytical procedures that exceed those normally associated with a remediation project. In addition, costs presented here are those that exceed the baseline investigations at a site. In other words, all sites have certain costs of sampling to determine the nature and extent of contamination.

These initial costs would be the same regardless of the technology chosen. A lack of cost and performance data is normal for innovative technologies at the pilot stage. That being the case, the goal of these case studies is to outline the information that is currently available. This information will hopefully provide a history of each site and the reasons for choosing phytoremediation as a treatment. In addition, these studies will provide a snapshot of the current status of these projects.

Aberdeen Proving Grounds-Edgewood Area J Fields Site

The Aberdeen Proving Grounds in Maryland began serving as a U.S. Army weapons testing facility in 1918. The installation is divided into two sections, the Edgewood Area and the Aberdeen Area, separated by the Bush River. Military weapons testing and past disposal activities over the years have caused extensive pollution throughout the soil and groundwater of the Proving Grounds. As a result, the entire Edgewood area of Aberdeen appears on the Superfund National Priority List (NPL).

Today the Department of Defence (DOD) and the Environmental Protection Agency (EPA) are jointly funding pilot scale applications of innovative treatment technologies around the facility. At the J Fields Site in the Edgewood Area, the EPA's Environmental Response Team (ERT) coordinated the planting of hybrid poplars over a shallow plume of chlorinated solvents in an effort to hydraulically contain the contaminants and treat the groundwater.

Site Design, Monitoring, and Goals

The J Fields Toxic Pits Site had been used for many years as an open pit burning facility for munitions and chemical agents. During this process, large volumes of various chlorinated solvents were discharged. As a result, there is a plume of chlorinated solvents located in the aquifer below the burning pits.

Table lists the contaminants of concern at the J Fields site and their concentrations in the groundwater. Concentrations of total VOCs in the groundwater range from less than 20,000 ìg/L in some areas to over 220,000 ìg/L in others.

Table. Contaminants of Concern at Aberdeen Proving Grounds J Fields Phytotremediation Site

Contaminant	Groundwater (µg/L)	Per cent (%)
1,1,2,2-tetrachloroethane (1122)	170,000	65.9
Trichloroethene (TCE)	61,000	23.7
Cis-1,2-dichloroethene (c-DCE)	13,000	5.0
Tetrachloroethene (PCE)	9,000	3.5
Trans-1,2-dichloroethene (t-DCE)	3,900	1.5
1,1,2-trichloroethene (TCA)	930	0.4

Personnel from Roy F. Weston, Inc. were contracted to assess the J Fields site and conduct treatment activities. Several technologies were considered for cleaning the soil and groundwater at the site. Soil washing, vapour extraction, and capping were considered for soils, and pump and treat and air sparging were considered for the groundwater. These technologies were eliminated from consideration for a number of reasons.

Technologies that involved a rigid installation design were eliminated because of a perched water table and the potential for unexploded bombs buried on-site. Pumping and treating the water would be difficult because of the high concentrations of contaminants and strict discharge regulations. In other words, the pump and treat system would need to remove high concentrations of contaminants from large volumes of groundwater, and then discharge the groundwater after it had been treated. Soil excavation was eliminated from consideration due to its high cost. After eliminating the other possibilities, project managers decided the J Fields site was a candidate for a pilot scale phytoremediation system. Applied Natural Sciences, Inc. was subcontracted to design and install the phytoremediation system. The phytoremediation strategy employed at the J Fields site began in September of 1995 with an assessment for phytotoxicity of on-site pollutants and to determine any nutrient deficiencies that would hinder tree growth.

In March and April of 1996, 183 hybrid poplars were purchased from a tree farm in Pennsylvania and planted over the areas of highest pollutant concentration around the leading edge of the plume, totaling about one acre of trees. A sweetgum tree was growing on-site prior to installation of the phytoremediation system, so it was left standing. It will be monitored along with the poplars. In order to promote growth down to the saturated zone, each tree was planted with a plastic pipe around its upper roots.

A long piece of rubber tubing was also added from the surface to the deeper roots in order to provide oxygen. A drainage system was installed in May 1996 to remove rainwater and therefore promote the plants' roots to seek groundwater. Since the Aberdeen project involves a new treatment strategy, extensive monitoring is taking place to determine the fates of the pollutants, the transpiration rates of the trees, and the best methods for monitoring phytoremediation sites.

The monitoring approaches are summarized in Table. The sampling design of the site involves collecting soils, transpiration gases, and tree tissues from the roots, shoots, stems, and leaves. Results will help determine the concentrations of contaminants and their metabolites along each step of the translocation pathway.

Table. Monitoring Approaches at the J Fields Site

Type of Analysis of Observation	Parameters Tested or Methods Used
Plant growth measurements and visual observations	Diamete,r height, health, pruning, replacement
Groundwater and vadose zone sampling and analysis	14 wells and 4 lysimeters to sample for VOCs, metals, and nutrients
Soil sampling and analysis	Biodegradation activity, VOCs metals
Tissue sampling and analysis	Degradation products, VOCs
Plant sap flow measurements	Correlate sap flow data to meteorological data
Transpirational gas sampling and analysis	Explore various methods

Eight monitoring wells were in place at the time of tree planting, and five additional wells were installed in November 1996. Two pairs of lysimeters were also installed on site. Tree sap flow rates are also being monitored in order to determine the pumping rates of the trees. An onsite weather monitor was used during sampling events to correlate tree evapotranspiration rates with weather fluctuations ranging from hourly to seasonal changes. Several different protocols were used for transpirational gas studies to investigate the most accurate and efficient monitoring methods.

All gas samples were collected in a 100 L Tedlar bag sealed over tree branches. Two of the sample collection media involved collection followed by analysis with a gas chromatograph (GC)/mass spectrometer(MS). One of these collection mediums was a Tennax/Carbon Molecular Sieve tube at collection rates ranging from 20-40 mL/min, and the other a 6 litre Summa canister. Perhaps the most accurate method involved a direct connection of Teflon tubing from the sample collection bag to a gas chromatograph (GC)/mass spectrometer(MS) quadrapole system.

This method allowed real time analysis of the transpiration gases. However, this method involves expensive monitoring equipment, so it probably will not be practical at most cleanup sites.

Cost

The trees cost about $80/tree to install. This works out to roughly $15,000 for installation of 183 trees. Costs of monitoring are highly varied due to the numerous monitoring techniques that have been employed at the site.

Performance to Date

Sap flow rate data indicates that on a daily scale, maximum flow occurs in the morning hours. In addition, increasing amounts of solar radiation seems to increase sap flow rates, as would be expected in a tree. Groundwater monitoring data from May of 1997 indicates that the trees are pumping large amounts of groundwater. Data indicates that there is roughly a 2 foot depression in the water table beneath the trees in comparison to data from April of 1996.

Tree tissue samples indicate the presence of trichloroacetic acid (TCAA), a breakdown product of TCE. This correlates with the results of Newman *et al.*, who also found TCAA in plant tissues in both axenic poplars cell cultures and hybrid poplar tissues in a greenhouse scale study. Site managers at Aberdeen are also finding that chlorinated solvents (TCE and 1,1,2,2-tetrachloroethane) are being evapotranspirated by the trees. To date, no mass balance studies have been performed to quantitatively determine the different fates of chlorinated solvents in this treatment system.

Future monitoring of the site will hopefully answer some of the questions about solvent fate. In order to accomplish this, additional types of monitoring will be employed, such as on-site infrared spectrometry and an onsite gas chromatograph/mass spectrometer. One other piece of noteworthy information is that the J Fields site experienced about 10% tree loss during the first year. While some of this loss was due to the transplant process, many trees were damaged by deer rutting.

In an attempt to keep deer away from the trees, the site managers hung bars of Ivory soap tied to string from the tree branches. The theory was that deer avoid the scent of humans, so they would think that there were humans nearby if they smelled soap. However, this did not completely deter the deer because there was still some rutting damage. Site managers plan to initiate some new strategies for the next fall.

Edward Sears Property

From the mid-1960's to the early 1990's, Edward Sears repackaged and sold expired paints, adhesives, paint thinners, and various military surplus materials out of his backyard in New Gretna, NJ. As a result, toxic materials were stored in leaky drums and containers on his property for many years. The soil and groundwater were contaminated with numerous hazardous wastes, including methylene chloride, tetrachloroethylene, trichloroethylene, trimethylbenzene, and xylene.

After his death, no one could be found responsible for the site or its clean-up, so On-Scene Coordinators (OSC) from EPA's Region II Removal Action Branch were called in to remove the leaking drums of hazardous materials, including off-spec. paints and solvents. Soil sampling indicated that two areas, 35 x 40 feet and 15 x 20 feet, were very heavily contaminated with solvents. These soils were removed to 8 feet below ground surface (just above the water table). Further excavation could not be achieved without pumping and treating large volumes of ground water.

The excavated area were backfilled with clean sand and the OSC activated the EPA's Environmental Response Team (ERT) of Edison, NJ to determine the extent of ground water and deep soil contamination. Using innovative, hydraulic push ground water sampling techniques, the ERT investigation revealed localized, but highly contaminated ground water. Based on this information, a limited number of monitoring wells were installed to determine vertical contaminant migration and to conduct aquifer tests necessary to evaluate pump and treat options. A test pilot for a pump and treat system with air stripping and activated carbon was then conducted.

The aquifer tests revealed a high yield aquifer, which required severe over pumping to create any substantial cone of influence around the pumping wells. Contaminants trapped in the silty-clay lens beneath the site would be difficult to extract in this manner because the transfer rate of contaminants into the groundwater is slow. As a result, large volumes of groundwater would need to be pumped to the surface for treatment, and this water would contain low concentrations of contaminant. Also, neighbors of the property would be disturbed by the noise created by a pump and treat system.

Based on these results, a pump and treat option would be expensive and inefficient for the Edward Sears site. Site managers then moved to a phytoremediation option. This site was judged as a potential candidate for a phytoremediation system due to the nature of the soils and groundwater. There is a highly permeable sand layer about 4-5 feet below ground surface, but below that exists a much less permeable layer of sand, silt, and clay from 5 to 18 feet bgs. This silt, sand, and clay layer acts as a semiconfining unit for water and contaminants percolating down towards an unconfined aquifer from 18 to 80 feet bgs. This unconfined aquifer is composed primarily of sand and is highly permeable.

The top of the aquifer is about 9 feet bgs, which lies in the less permeable sand, silt, and clay layer. The top of the aquifer is relatively shallow and most of the contamination is confined from 5 to 18 feet bgs, so managers decided to plant hybrid poplars in order to prevent further migration of the contaminants and ultimately remove the contaminants from the groundwater. Samples were taken from temporary well points throughout the site. Data from these sampling efforts indicated TCE concentrations in the groundwater ranged from 0 to 390 ppb.

Most of the TCE was concentrated into a small area on-site. Seven monitoring wells were installed based on the information obtained from the temporary well points. Data from Monitoring Well 1 can be found in Tables 17 and 18. Keep in mind that these tables only provide information from Monitoring Well 1, and there were a total of 7 wells. Monitoring Well 1 was installed in the area of highest TCE contamination.

There was little or undetectable TCE found in the groundwater samples from the other 6 wells. That does not mean that there were no contaminants found in those wells. Recall that the site was polluted with a wide variety of organic chemicals and metals due to the storage practices of Mr. Edward Sears. However, since this report focuses mainly on TCE, only data on TCE from Monitoring Well 1 was included.

Table: Concentrations of TCE Sampled from Groundwater in Monitoring Well 1 on Edward Sears Property

Sampling Date	TCE Concentration (ug/L)
12/8/95	28
8/8/96	1.2
8/19/96	2.3

Table: Concentrations of TCE in Soil Sample During Monitoring Well Installation of Monitoring Well 1 at Edward Sears Property

Feet bgs	1	4	6	8	10	11	12	14	16	18	20
ug/kg	130	18,000	540	270	120	140	48	17	35	180	8
Feed bgs	21	27	32	37	42	47	52	57	62	67	72
ug/kg	120	130	6	62	3	65	100	16	6	5	6
Feet bgs	75	77	82	87							
ug/kg	52	U	4	7							

Site Design, Monitoring, and Goals[SH]

Roy F. Weston, Inc., under the Response Engineering and Analytical Contract (REAC), was tasked by the ERT to conduct a pilot phytoremediation test at the Sears site. The test is being conducted to determine whether hybrid poplar trees can be used to reduce soil and groundwater VOC contamination levels in the planted area and to prevent further offsite migration of contaminated groundwater.

In October and November of 1996, the site was cleared of debris and a 4-inch clay layer was placed approximately 1 foot bgs to prevent penetration of rainwater into the upper root zone, thus promoting root growth into the underlying aquifer. This was followed by the replacement and grading of the native surface soil. Thomas Consultants of Cincinnati, OH were subcontracted to layout the phytoremediation design. In December 1996, one hundred and eighteen hybrid poplar saplings were planted by ERT, REAC and Thomas Consultants personnel in a plot approximately one third of an acre in size.

The trees were planted 10 feet apart on the axis running from north to south and 12.5 feet apart on the east-west axis. The trees at Ed Sears were planted using a process called deep rooting. In deep rooting, the roughly 12 foot trees were buried nine feet under the ground so that only about 2-3 feet remained on the surface. Deep rooting the trees involved drilling 12 inch diameter holes to a depth of 13 feet. These holes were then back filled to 5 feet below ground surface with amendments such as peat moss, sand, limestone, and phosphate fertilizer.

This backfill was installed to provide nutrients to the roots as they penetrated down through the soils. Waxed cardboard cylinders 12 inches in diameter and four feet long were installed to promote root growth down into the groundwater. These barriers settled about a foot into the planting holes, so 5 gallon buckets with the bottoms cut out were placed on top of the cylinders to create a 5 foot bgs root barrier. The trees were placed in the cylinders and the remaining five feet to surface was filled with clays removed during the boring process.

There were about 90 extra poplars left after the deep rooting was completed. These extra trees were planted along the boundary of the site to the north, west, and east sides of the site. These trees were only planted to a depth of 3 feet, or shallow rooted.

The purpose of the shallow rooted trees was to prevent rainwater infiltration from off-site and to serve as a source of replacement trees in the event that there was a loss of some deep rooted trees. These trees were planted very close together (about 3 feet apart) under the assumption that natural thinning would take place over subsequent growing seasons.

A surface water control system was then installed by planting grasses over the entire site. These grasses came from commercially available seeds purchased from a lawn and garden store. ERT is conducting an ongoing maintenance and monitoring programme at Ed Sears.

Monitoring of the site includes periodic sampling of groundwater, soils, soil gas, plant tissue, evapotranspiration gas. Continued growth measurements will also be made as the trees mature. In the fall, the surface water control system will be replaced due to a summer drought that killed much of the grass.

Site maintenance also involves the prevention of deer and insect damage. Bars of soap were hung from the trees to deter deer from rubbing their antlers on the trees. Some damage was inflicted by an insect larva known as the poplar leaf caterpillar.

This caterpillar lives on poplar trees and makes its cocoon by rolling itself in a poplar leaf. A spray containing Bacillus thuringesis, a bacteria which produces toxins that are specific to various insects, was applied to the site. This spray has been effective in killing most of the caterpillars that were living on the trees.

Cost

Total cost of installation of the 118 deep rooted and 90 shallow rooted trees was about $25,000. Additionally, installation of the surface water control system and one year of on-site maintenance totaled about $15,000.

Performance to Date

The trees have been in the ground for less than one growing season, so as of now there is very little performance data available. Some sampling of evapotranspiration gas was conducted by placing Tedlar bags over entire trees. Data from these air samples suggests that the trees are evapotranspirating some VOC's. However, the VOC concentration in the Tedlar bags matches the background concentrations of VOCs in control samples. This could be due to VOCs volatilizing from the soils, or it could be due to evapotranspirated VOCs that may have gotten into the control samples. Future sampling designs will attempt to determine accurate background VOCs.

Additionally, there is some tree growth data for the trees. They have grown about 30 inches above ground since planting. Figures 4 and 5 are photographs of the trees at Edward Sears taken in July 1997 showing their size after about 7 months of growing on site. Site managers plan to sacrifice one tree either after or during the next growing season to determine the extent of root growth.

Carswell Air Force Base

In Ft. Worth, TX, the U.S. Air Force planted Eastern cottonwoods (Populus deltoides) to investigate the ability of these trees to control and degrade a plume of TCE in a shallow aquifer. The plume is located near Air Force Plant 4 at the Naval Air Station Ft. Worth, which was formerly known as the Carswell Air Force Base.

The initial funding and much of the ongoing support for this project was provided by the Environmental Security Technology Certification Programme (ESTCP), a division of the Department of Defence. The Carswell site was chosen as an EPA Superfund Innovative Technology Evaluation (SITE) project in 1996.

Site Design, Monitoring, and Goals

Greg Harvey of the U.S. Air Force Acquisition and Environmental Management Restoration Division and Steve Rock of the US EPA National Risk Management Research Laboratory carried out the design and implementation of the phytoremediation strategy at Carswell. In April of 1996, the USAF planted 660 cottonwoods in an effort to contain and remediate a plume of dissolved TCE located in a shallow alluvial aquifer (6 to 11 feet below grade). The species P. deltoides was chosen over a hybridized species of poplar because it is indigenous to the region.

Therefore it has proven its ability to withstand the Texas climate, local pathogens, and other localized variables that may affect tree growth and health. Two sizes of trees were planted: whips and five gallon buckets. The whips were about the thickness of one's thumb and were about eighteen inches long at planting. The whips were planted so that about two inches remained above ground and the rest of the tree was below ground to take root.

The five gallon bucket trees were about an inch in diameter and seven feet tall when planted. The five gallon bucket trees were estimated to have about twice as much leaf mass as the whips when planted, so they were expected to have higher evapotranspiration rates. The layout for the project involved planting a separate plot of trees for the whips and the 5 gallon buckets, with both plots perpendicular to the contaminant plume.

The plume is moving to the south and east, so the plots were laid out on a north and east axis. The whips section was planted to the north and west of the 5 gallon buckets, so that the plume would first travel through the root zone of the whips and then through the root zone of the 5 gallon buckets. A control area with monitoring wells was placed to the north and west of the whips, and another in between the whips and the five gallon buckets, along with monitoring wells throughout the treatment site.

These control areas enable data to be collected on the amount of contaminant that enters each of the treatment areas (whips and five gallon buckets), so that a comparison of the performance of each type of tree can be made. Figure contains the basic layout of this site. One unique aspect of Carswell is that there is already a mature cottonwood growing on the site. This 70 foot tall tree is located just south and east of the planting area on the other side of a cart path.

Groundwater monitoring wells were installed around this tree, and it will be sampled in a similar manner to the planted cottonwoods to see how well a mature tree functions in this phytoremediation system. Data from the first three groundwater samplings that have taken place indicate that the wells near the mature tree have lower concentrations of TCE. This observation is promising because it may be indicating that the older tree is treating the TCE in the groundwater at a higher rate than the younger trees. However, these observations are only speculative at this time.

Costs

Some rough estimates of cost for the Carswell site were provided by Mr. Greg Harvey. These estimates can be found in Table . Since this site involves an innovative treatment technology, these costs are substantially inflated due to the heavy monitoring taking place at the site. Also, there are no long term projected costs available or total costs for the project available because the time involved in remediating the site is uncertain.

In addition to the costs in Table $200,000 will be spent for extensive site monitoring that would not normally be associated with a phytoremediation system, so this amount was not included in the cost estimates.

Table. Rough Estimate Costs of Phytoremediation at Carswell Air Force Base

Activity	Estimated Cost
Wholesale cost of trees (does not include delivery or installation costs)	$8/tree for five-gallon bucket tree $0.20/tree for whips
29 wells (including surveying, drilling, and testing	$200,00
Subsurface fine biomass	$60,000

Performance to Date

Evapotranspiration rates at Carswell for May 13 and 15 and June 10, 11, and 12 of 1997 have been determined. Unfortunately, no quantitative evapotranspiration data was available to include in this report. Qualitatively, both types of trees were capable of evapotranspiring TCE, and the 5-gallon trees are evapotranspiring more water than the whips. This was to be expected because of the greater total surface area of the 5-gallon trees' leaves. In addition, the transpiration rates were generally higher in June than May, which is likely due to a combination of warmer weather and more fully developed leaves.

There also appeared to be a midday decline in transpiration during June, indicating that the plants were experiencing water stress during the hottest part of the day in the summer months. In other words, the water demand for the tree exceeded the supply during that time. There was also a notable difference in transpiration rates between days in June, with cloudier days resulting in lower transpiration rates. In addition to evapotranspiration information, some tree growth data has also been collected. In 16 months the whips have grown about 20 feet.

and the 5 gallon bucket trees have grown faster than the whips. Groundwater samples were collected from the 29 monitoring wells and analysed on three occasions to date. Concentrations of TCE, cis-DCE, and trans-DCE, and vinyl chloride were determined from these samples.

They ranged from 2 to 930 ug/L TCE in the groundwater, with most samples falling in the 500-600 ug/L range. Average concentrations of the contaminants on the three sampling dates are provided in Table, with the exception of vinyl chloride. Vinyl chloride was only detectable in a handful of samples and generally in low levels, so an average concentration was not determined.

Table. Average Concentrations of TCE, cis-DCE, and trans-DCE at Carswell

Contaminant	Average Concentration (ug/L)		
	December 1996	May 1997	July 1997
TCE	610	570	550
cis-	130	140	170
trans-DCE	4	2	4

Some analytical work has been done on the tree tissues at the site, but this type of information is still in the early stages of collection. Data from November of 1996 indicated a TCE signature in the whips that were planted over an area where the groundwater was the shallowest. This indicates that the young trees were capable of evapotranspirating TCE after just one growing season.

Now that the trees have been on site for more than an entire growing season, site managers at Carswell plan to increase monitoring at the site to include a whole suite of water, soil, air, and tree tissue sample analysis.

Some of the more unique data they plan to collect (in relation to the other case study sites) are analyses of microbial populations and assays of TCE degrading enzymes in the trees.

The Future of Phytoremediation

The use of phytoremediation will most likely increase in the near future. As an example, Table provides an analysis of the phytoremediation market for organics in groundwater through the year 2005. Assuming these predictions are accurate, phytoremediation will likely grow into a substantial market over the next decade.

Of course, the results of the three case studies and other similar studies currently being designed and implemented may have an effect on these predictions.

Table. Estimated U.S. Phytoremediation Market Share for Organics in Groundwater

Market	Cost in Millions of Dollars		
	1997	2000	2005
Total segment	2,600	2,600	2,600
Phytoremediation	2-3	6-12	20-45

Global Warming

Table outlines some of the research needs that the Remediation Technology Development Forum (RTDF) for Phytoremediation of Organics believes are needed in the near future. Because phytoremediation is such a new technology, most of the needs outlined in this list are still at a fundamental level. For example, very little is known about the mass balance of fates for many pollutants within plants and the degree that the rhizosphere can increase degradation.

Basic Research Needs for Phytoremediation:

- More basic data is needed on the phytoremediation process
  - Validation of rhizosphere effects
  - Determine the fate of a contaminant in phytoremediation systems
  - Determine factors that affect mass balance, such as species, climate, soils, etc.
- Determine acceptable endpoints for phytoremediation
- Selecting a contaminant and site that will effectively convince regulators that phytoremediation is a valid Technology

Once many of the basic questions have been answered, research will need to focus on optimization of phytoremediation for different clean-up situations and goals. This can be accomplished through a number of mechanisms, from complex studies such as genetic engineering to simply finding better ways to choose native plants. One way to accomplish this is to choose native plants that produce high concentrations of enzymes that are known to degrade a site's contaminants of concern.

Table outlines some of the enzymes known to work in phytoremediation systems and some of the plants that contain these enzymes. Researchers need to screen more plants for these and other enzymes potentially useful to phytoremediation in order to create a catalogue of plants. Today, trees such as poplars are commonly used in phytoremediation schemes because they can grow in a wide variety of climates and they are known to have high evapotranspiration rates. However, there may be plants that are better than poplars at degrading specific pollutants. For example, cypress trees and rice plants were found to contain much higher concentrations of dehalogenase than poplars, so in theory they are better equipped to degrade TCE than a poplar tree (Laura Carreira, personal communication). In addition, there may be certain trees native to an area that would be much heartier under local conditions and more resistant to local pathogens and parasites, so it is often beneficial to chose a native plant when designing a phytoremediation scheme.

Choosing native species and screening them for their ability to metabolize specific contaminants may be the key to optimizing phytoremediation in the future. Currently, there is speculation as to how well phytoremediation works and under what conditions it will be useful. Many of these questions will be answered, at least in part, by pilot scale projects such as Aberdeen, Carswell, and Edward Sears, as well as the numerous other types of phytoremediation projects underway around the country. Early results from these current sites are promising, so it appears that phytoremediation will be an effective tool for cleaning hazardous waste in the future.

Chapter 6

Solid Waste Management

The Office of Sustainable Development and the Noise Control Office to review the commercial waste and recycling system and report back on potential changes to address not only complaints about nighttime collection but broader environmental and economic issues. City Council requested that a discussion paper be developed providing an analysis of various alternatives and a process for proceeding forward.

That discussion paper is hereby submitted as an attachment to this resolution, Whereas, by Ordinance 171067, in April 1997, the Council set the City's waste recovery goals at 54% in 2000 and 60% in 2005. Portland met the 54% goal but has not reached the 60% goal set for 2005, Whereas, recovery rates have leveled off since 2000, but the amount of garbage disposed of in landfills continues to grow, increasing more than 7% in 2004 alone, Whereas, these trends, if they continue, will make it very difficult to achieve higher performance and long-term sustainability of the solid waste and recycling system, Whereas, the greatest opportunities to reduce the environmental impacts associated with the generation and management of waste occur before those resources enter the solid waste system.

Instead of focusing on avoided disposal, waste prevention in all phases of production and consumption should be the prevailing management strategy, Whereas, Extended Producer Responsibility is a policy approach in which producers assume responsibility for the management of post-consumer products, so that those who produce and use products bear the costs of recycling and proper disposal.

When brand owners are responsible for ensuring their products are recycled and the associated costs are included in the product price, there is a strong incentive for producers to design, and consumers to purchase, goods that are more durable, easier to recycle, and less toxic, Whereas, there are significant economic, social and environmental benefits associated with greater recovery and waste prevention including energy savings, reduced greenhouse gas emissions, air quality improvements, resource conservation, job growth and economic development opportunities, Whereas, the City's efforts to initiate a new voluntary food waste recovery programme in the commercial sector have produced limited results due to economic barriers and a lack of incentives inherent within the current system of regulations and collection programs, Whereas, changes must be made to the programs, policies and regulatory framework for solid waste and recycling collection and disposal to produce necessary gains in waste prevention and recovery and lessen the environmental and human health impacts of the system as a whole, Whereas, the Sustainable City Principles adopted in 1994, require that the City promote a sustainable future and follow a set of comprehensive principles incorporating environmental quality, social equity and economic vitality, Whereas, the Local Action Plan on Global Warming adopted by the City of Portland and Multnomah County in 2001 directs the City to "promote solid waste management practices that reduce greenhouse gas emissions and promote community understanding of the relationship between solid waste reduction and global climate change," Whereas, the City in its own operations should lead by example, Whereas, it is expected that any changes in policies, programs or regulatory framework for the solid waste and recycling system will significantly influence the noise issues related to nighttime collection. Further action to resolve the noise issue should be considered in light of any proposed system changes,

POLICY

Therefore, be it resolved, that the Council adopts the following goals:

- Promote sustainability of the solid waste and recycling system that includes maximum efficiency, equity and economic vitality, improved worker safety and reduced environmental and human health impacts over the entire life cycle of the materials, and
- Minimize the impact of harmful wastes by targeting toxicity and reducing greenhouse gases emissions,
- Reduce per capita waste generation below 2005 levels by the year 2015, and
- Maximize recovery of all waste with a target of 75% by the year 2015 and promote highest value use of the recovered materials.

Therefore, be it further resolved, that the Office of Sustainable Development is hereby directed to prepare and submit to the Council a Solid Waste Management Plan and conduct a planning process to solicit input from the public and stakeholders in developing proposed solutions. In the planning process, a variety of strategies shall be considered, including Extended Producer Responsibility policies. The Plan shall address ways to reach new recovery and waste prevention goals, minimize harmful wastes and improve the performance and long-term sustainability of the solid waste and recycling system. The Plan shall include recommendations for both the residential and commercial sectors as well as City government operations.

Therefore, be it further resolved, that this resolution constitutes non-binding city policy, in directing the development of further policies to be brought back before the Council. City Council directed the Office of Sustainable Development (OSD) and the Noise Control Office to review the commercial waste and recycling system and report back on potential changes. While initially driven by the issue of noise associated with nighttime collection, the resulting analysis revealed several additional concerns:

- Recycling rates in the city have leveled off over the past five years at about 53%, falling short of the 60% goal for 2005.
- Waste produced by businesses in Portland accounts for 75% of the total. While the residential sector produces less waste than the commercial sector, there are important synergies between the two in recycling behaviour and collection practices.
- While the recycling rate is relatively flat, total waste generation is increasing and more waste is going to landfill disposal. In 2004, overall waste generation was up 3.3% over the previous year and solid waste disposed grew by 7.4%. If these trends continue, recycling performance will fall further and further behind the 60% goal.
- A voluntary commercial food waste collection programme has yielded limited results due to economic barriers within the existing system.

These developments make it clear that there are broader issues at hand than the noise concerns which prompted the review. Any reform of the garbage and recycling collection system must address the structural barriers that are preventing higher performance and long-term sustainability.

Current Commercial System

Unlike most other cities and counties in the metropolitan region, Portland's commercial garbage and recycling system is defined by free market conditions and competitive pricing of services. While this competitive system provides an opportunity for choice by local firms, and may provide lower prices for some of the City's larger businesses, it does not give adequate incentives to either businesses or haulers to reduce garbage or recycle more waste, and is inequitable to smaller businesses. A key weakness of Portland's existing commercial collection system is that it does not provide a reliable means for compensating haulers for the cost impacts of new recycling requirements. Consequently, haulers are not rewarded for promoting recycling and businesses may not have the means or the motivation to expand their recycling systems.

OSD staff believes that all the gains to be expected from the commercial sector under the current system of regulations and collection programs have been achieved. Unfortunately, a detailed look at commercial garbage reveals that more than half (60%) of what Portland businesses still throw away is material they could recycle.

A More Sustainable Approach

The review of the commercial system also spurred a broader discussion about what goals should be driving any proposed reform. Members of City Council expressed concerns over the broader environmental and human health impacts of waste generation and collection.

In addition to achieving higher recycling levels, the system should be moving towards reductions in energy use, greenhouse gas and other toxic emissions, and in the amount of waste produced overall. In short, a more sustainable approach is needed to define success and guide the evolution of the system, one that is predicated not just on greater recycling but more aggressive waste prevention as well.

Such an approach is consistent with the Sustainable City Principles adopted by Council in 1994 that require the City to promote a sustainable future and follow a set of comprehensive principles incorporating environmental quality, social equity and economic vitality.

These goals are mirrored in the Local Action Plan on Global Warming adopted by the City and Multnomah County in 2001, which directs the City to "promote solid waste management practices that reduce greenhouse gas emissions and promote community understanding of the relationship between solid waste reduction and global climate change."

The concept of Extended Producer Responsibility is an emerging product stewardship approach to achieving greater waste prevention that has been utilized by local and state governments. In general terms, the approach calls for shifting the waste management burden from local governments to producers and manufacturers for the collection, recycling and disposal of their products. When manufacturers must absorb the end-of-life costs of their products and these costs are reflected in product pricing, manufacturers have an incentive to design, and consumers to purchase, goods that are more durable, easier to recycle and less toxic.

Environmental and Human Health Benefits

Research into the benefits of greater recycling and waste prevention presents some compelling evidence but also points out some important considerations. First, it is important to note that waste prevention activities capture most of the same benefits as recycling but in greater magnitude by virtue of the avoided resource use in extraction, processing, transportation and disposal. Also, where and how the recovered materials are used makes a difference. For example, recovered paper used to produce newsprint has a much greater environmental benefit than using the same recovered paper as animal bedding.

Some of the gains specific to recycling include:

- *Energy savings*: It takes 95% less energy to recycle aluminum than it does to make it from raw ore. Making recycled steel saves 60%; recycled newspaper 40%; recycled plastics 70% and recycled glass 40%.
- *Global warming*: Portland's current recycling rate is a significant factor in reducing CO_2 emissions and reaching global warming goals for the city and the region. For example, every ton of food waste diverted to composting prevents approximately one ton of CO_2 equivalent in greenhouse gas emissions from the landfill.
- *Air quality improvements*: The Metro area recycled 437,000 tons of paper in 2003, which is equivalent to taking 200,237 cars off the road for one year in the reduction of carbon monoxide, hydrocarbons and nitrogen oxides.
- *Resource conservation*: In 2003, our region recycled $79 million worth of materials that otherwise would have been thrown away. The Portland area recycles enough paper each year to save about 8.2 million trees, equivalent to the number of trees in eight Forest Parks.
- *Job growth*: Recycling provides a large number of jobs that generally pay above the national average wage. Sending 10,000 tons of waste to the landfill supports 6 jobs, while recycling the same amount supports 36 jobs. In the Portland area, 1,000 people work in the recycling industry.
- *Economic development*: Demand for recycled materials has never been greater and, in many cases, exceeds the supply currently

provided by the public. New business and product opportunities associated with recycled materials will add to the growth of the sustainable industries sector of our local economy.

Options for Change

If we are to meet the City's recycling goals and take steps towards more aggressive waste prevention, specific changes must be introduced to improve the performance of the system. Attention should be given not only to the commercial sector, but to the residential sector and city government operations as well.

Commercial

For the commercial programme, there are several options to consider:

- Continue current commercial structure, yet require additional business recycling.
- Establish a uniform rate structure for various levels of service, and require additional recycling.
- Develop a commercial franchise system based on the existing haulers. Include additional recycling requirements and waste prevention goals as part of the franchise.
- Develop a commercial franchise system with fewer haulers that are chosen through a competitive process. Include additional recycling requirements and waste prevention goals as part of the franchise.

There are many tradeoffs between:

- Allowing businesses the freedom to choose their hauler and negotiate price.
- Regulating services and rates that result in new more effective programs and increased waste reduction and recycling.

It has been widely assumed that if the City set rates to compensate haulers for recycling and other mandatory services, then overall costs to businesses would rise. Research conducted by OSD indicates that may not always be the case and that small businesses in particular may be paying more under the current system than their counterparts in neighboring suburban communities with franchised service.

To gather specific information on these tradeoffs from the perspective of various stakeholders, OSD held a series of professionally facilitated meetings. The primary and most obvious finding: There is no agreement among commercial haulers, the business community, environmental organizations or neighborhood associations on the preferred option for improving the commercial collection system.

Residential

Changes may also be warranted in the residential curbside collection programme to see greater overall sustainability in the system. With hauling services franchised for this sector, the advantage of a rate-setting mechanism that provides for cost recovery allows for the implementation of a wide range of system improvements. Some examples under consideration include:

- Use of alternative or biofuels and retro-fitting of trucks to reduce emissions associated with collection.
- Use of roll carts and automated collection to increase convenience and participation in curbside recycling and yard debris composting and reduce worker injuries.
- Expansion of the curbside programme to include food scraps in with the yard debris for composting.

City Operations

It is important that the City lead by example in its own recycling and waste prevention activities. Opportunities exist in the full range of purchase, reuse or recycling and disposal practices. Several relevant initiatives are underway including the Sustainable Procurement and Sustainable Paper Use policies, a draft Toxics Reduction Strategy and consolidation and redevelopment of waste hauling contracts for city facilities. An expansion of these efforts and identification and implementation of other prospects is an appropriate addition to the overall strategy.

OSD recommends Council adopt a resolution directing OSD to undertake the following actions:

- Produce a new Solid Waste Management Plan, including a strong focus on sustainability, waste prevention, commercial solid waste collection and recovery.
- Develop a planning process with the Solid Waste Advisory Committee to include all stakeholders and review all options.
- Engage the broad community and specific stakeholders to identify issues and options for new policies, programs and systems. These options should include actions for the City to take affecting all sectors of the community and specific actions related to preventing waste within City government operations.
- Develop a draft plan and circulate for public review, discussion and comment.
- Integrate comments and recommendations from the public review, develop the final plan and take to City Council for approval.

To address the noise issue associated with nighttime collection, it is expected that any changes in policies, programs or regulatory framework for the solid waste and recycling system will significantly influence the collection practices that have been the source of complaints. OSD recommends that anywhere action to resolve the noise issue be considered in light of proposed system chaı ges coming from the Solid Waste Management Plan process.

MUNICIPAL SOLID WASTE MANAGEMENT

The problem of solid waste management (SWM) is acquiring an alarming dimension in India. The swift industrialization and urbanization has resulted change in the life style, because of which the quantity of solid waste generated has increased significantly and its characteristics have changed. Lack of financial resources, institutional weakness, improper choice of technology and public apathy towards SWM has made this service far from satisfactory.

In this context an investigation has been undertaken to study the present scenario of SWM in Raichur city of Karnataka. The entire city is divided into three sanitary zones for effective management of MSW covering all 35 wards. The samples were collected from all the three zones for characterization. The Raichur urban local body (ULB) operates to collect waste by door-todoor collection and containerized storage system.

The dumper placers are used to transport the waste to the dumping site. Auto tippers are used for primary collection in slum zones and tricycles are used in other parts. The disposal site is near Eklaspur village which is 6 km away from the city. At present solid waste in not segregated at the source and open dumping is practiced. However, both these practices are not complying with the guidelines laid by MSW (Handling and Management) Rules 2000.

The average per capita per day generation of solid waste was found to be 278 gms. A thorough study of all the components of municipal solid waste management was carried out and scientific management of MSW is suggested. In this paper the attempt have been made to proper management of municipal solid waste in Raichur city.

Municipal solid waste management (MSWM) is one of the major environmental problems of Indian cities. Improper management of municipal solid waste (MSW) causes hazards to inhabitants. Various studies reveal that about 90% of MSW is disposed of unscientifically in open dumps and landfills, creating problems to public health and the environment. Development of sprawling urban area and rapid industrialization has increased the quantity as well as diverse nature of wastes, which need proper handling and treatment. India has a mixed demographic profile with about 307 million (about 30%) of the total population living in urban area and by 2011 it will be 395 million. The present system of SWM in most of the urban areas are highly unsatisfactory due to limited finances, inadequate services and coupled with pubic apathy towards the same. The solid waste management is one of the essential service and it is an obligatory duty of municipal bodies to arrange for daily street cleaning, collection, transportation, processing and disposal of waste.

Improper handling and disposal practices of solid wastes continue to be a serious problem. Indian cities generate on an average of 300-400 gms/capita/day solid waste and of which only 60-80% of the waste is collected on daily basis and rest of the waste is left to decay on the streets, roads, drains etc, which attracts vector transmitting diseases. Technological development, globalization and population growth have accelerated the dynamics of the urbanization process in developing countries. In India alone, the urban population has increased from 11% in 1901 to 26% in 2001. The rapid growth rates of many cities, combined with their huge population base, has left many Indian cities deficient in infrastructure services like water supply, sewerage and solid waste management. The poor management of SW in urban area is putting public to serious health risks. Hence it is therefore to initiate concerned effort for proper scientific management of MSW.

STUDY AREA

Raichur is a district headquarters in Karnataka, India and is situated between two rivers Krishna and Tungabhadra at 150 12l N latitude and 760 21l E longitude which is popularly known as "DO-AABA". It receives annual rainfall of 638mm.The population of the town is about 2,62,500 and floati..g population of about 20,000.It is having good connectivity by road and rail.

It is also an important educational, administrative and trade centre in the Hyderabad-Karnataka, region. It is housing a very few industries within municipal limit. The city municipal council extends over an area of 60.0 sq kms. The entire area is divided into three sanitary zones for effective management of MSW covering 35 wards.

SOLID WASTE MANAGEMENT PRACTICE IN RAICHUR- PRESENT SCENARIO

Like Indian typical cities in the Raichur municipality the majority of the waste generated is from domestic sector followed by commercial establishments, markets, other public places and from open areas. At present about 73 tonnes of waste is generated per day. The per capita solid waste generation rate between 250 to 300 gms per day is observed during normal day.

The solid waste management of all 35 wards is handled through public participation by including self help groups and private contractors. However the segregation of waste at the source is not practiced. Unsegregated waste is collected by tricycles from the sources and carried to the secondary storage containers of 3 and 4cum capacity. Depending upon the waste generated and command area, containers are placed in different wards.

At present there are 60 containers and 5 dumper placers supported by 15 push carts, 80 tricycles and 7 auto tippers for primary collection. However there are still some areas where public throw their waste at their will on streets, roads etc. To facilitate the collection of these waste 3 tractors are used. Finally collected waste in containers is transported by the dumper placers to the open disposal site near Eklaspur village which is 6 kms away from the city. Table shows the solid waste generation of the Raichur municipality.

The physical components of MSW were analysed from various zones of Raichur city including the biodegradable and recyclables. Physical characteristics of the representative samples were studied in terms of there weight density, moisture content and are represented in Table.

Table. Source Wise Solid Waste Generation of the Raichur Municipality

Sl No	Sources	Total No	Unit quantity of waste per source in kg.	Total quantity of waste generated kg per day
1	Non slum house holds	26023	1.5	39034.5
2	Slum house holds	21126	1.0	21126
3	Commercial shops	3434	1.5	5151
4	Major Hotels and Resorts	12	15	180
5	Small Hotels	33	20	330
6	Markets	01	200	200
7	Choultries	23	10	230
8	Hostels and Institutions	99	5	495
9	Boarding and Lodging	15	15	225
10	Vegetable Shops	20	5	100
11	Meat Shops	136	5	680
12	Municipal Waste from Industries	149	10	1490

13	Theaters	12	4	60
14	Wastes from floating population			1000
15	Waste from street sweeping			2000
		Total:		72301

The average moisture content and density was fount to be 28.902% and 166.72 kg/m3 respectively. The Raichur MSW comprised of about 74% biodegradable waste, about 1.5% recyclables and 24.5% other waste which is shown in Figure.

Table. Constituents and Properties of Raichur MSW

Sl No	Components	Percentage by mass	Average Moisture content	Dry mass in kg	Volume M^3	Typical Density Kg/m^3
1	Food and vegetable waste	33.21	72	10.170	1.145	290
2	Paper	13.39	8	13.540	1.575	85
3	Plastic	7.10	4	6.350	1.109	65
4	Textile	2.65	9	2.410	0.407	65
5	Rubber	1.85	3	1.750	0.142	130
6	Leather	1.35	12	1.180	0.0845	160
7	Wood	1.98	18	1.200	0.0825	240
8	Dust, ash, brick	7.86	10	6.810	0.163	480
9	Glass	1.6	4	1.568	0.082	195
10	Rags and mix waste	29.01	6	26.100	1.208	240
	Total	100.00		71.098	5.988	

DESIGN OF BIO-COMPOST PLANT

A land area 24.18 acres is available for disposal. As about 74% of the waste generated is organic it is viable to have the biocompost plant as eco-friendly and sustainable method of disposal. The proposed biocompost plant is designed for the estimated population of 5,51,693 with 20 years of design life period. The estimated quantity of waste to be handled is 264.70 tonnes per day. The windrows are designed on the basis of 50 tonnes per day in the starting year. The windrow platforms are sized to accommodate the dumping area 4m wide, windrow width of 6m and 4m for I,II,III week and IV,V,VI week respectively. Clearance width of 3m for turning of windrows and passage of 4m wide for shifting the garbage from one windrow to next windrow for the transport Direct Seedingof digested garbage.

Platform is provided with 2% slope to ensure collection of leachate or surface run off through the lined drains provided all around the platform leading to leachate collection sump. The overall size of the plat form is 73m × 53m.

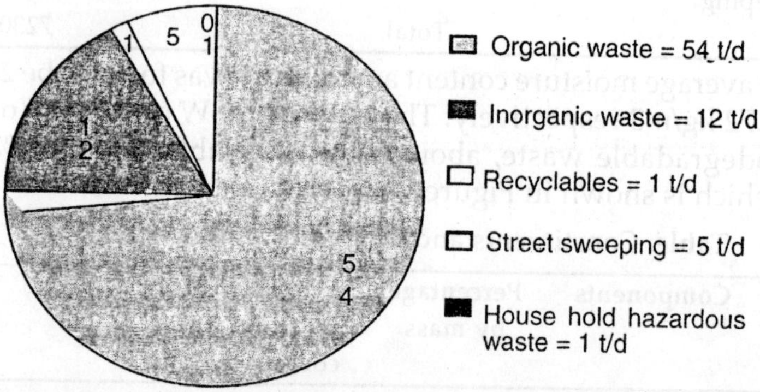

Fig. Composition of Raichur Municipal Solid Waste

In this work a detailed study of collection, storage, transport and disposal practices was conducted for Raichur city. An improvement to existing system has been proposed to meet the MSW (H&M) rules 2000. The site has been planned as an integrated facility for compost plant and sanitary landfill for a design period of 20 years. In order to build up institutional capabilities it is necessary to harness and integrate the role of other emerging actors in the field of SWM. Private sector, NGOs and rag pickers are to be brought in to the institutional framework for effective management of MSW.

DECENTRALIZED SOLID WASTE MANAGEMENT

Cleanliness is a major factor that influences development of any city or town, which is otherwise hampered due to improper collection and disposal of solid waste. Managing solid waste is one of the major tasks for the ULBs. Solid waste management is a continuous process and if unchecked can not only be a health hazard but will also impart multidimensional threats, includes serious detrimental, environmental, social, and economic impacts. For instance, improper disposal of waste had led to outbreak of bubonic Plague in 1994 in Surat (Gujarat). The trend of unsustainable patterns of production and consumption is increasing the quantity of the waste and the amount will increase four to fivefold by the year 2025. The only solution is to adopt scientific and economical action plan for solid waste management. This should integrate various practices for treatment and disposal involving different stakeholders for effective implementation. Decentralized solid waste management brings in effectiveness, The various stakeholders involve:

- Citizens/ residents
- Service provider - NGOs/ Private contractors
- ULB

WHAT LAW SAYS?

As per Supreme Court Guidelines for scientific management of solid waste called Municipal Solid Waste (Management and Handling) Rules 2000, SWM is one of the obligatory functions of the ULB. Now government of Karnataka has prepared State Policy complying with MSW rules and relevant tender documents, guiding the ULBs to follow a standard method for segregation, primary collection, secondary storage, transportation, processing and disposal of municipal solid waste. Community should be involved actively in various ULB's activities as per 74th Constitutional Amendment Act.

Advantages to the Residents

- Creates employment in the neighbourhood
- Allows quality surveillance of the waste management system by the residents
- Adoptable and sustainable process
- Decreases the problems of the malfunctioning municipal services

Advantages to the ULB

- Diverts organic waste from municipal solid waste stream at the local level thereby cutting down the transportation charges for the ULBs.
- Enhances environmental awareness
- Brings in responsibility among the community towards the waste generated by them.
- As per the MSW rules 2000, SWM is one of the obligatory function of the ULBs, there by reduces the municipality's responsibility of primary collection.

How to Involve Residents?

ULBs should conduct awareness programmes motivating the residents to establish their own system for SWM. They should look forward to involve residents in the form of RWAs for effective implementation of SWM action plan. ULBs should encourage RWAs in adopting their own system for SWM by providing financial and technical support.

Process Involved in decentralized Solid waste management system (DSWM)

- Source segregation of waste into wet and dry.
- Door to door waste collection
- Segregation of recyclables from the dry waste and selling the same by the retriever
- Composting of the wet waste
- Storage and transportation of dry waste into municipal trucks for landfill
- Preparing leaf mulch from the garden and street sweeping waste
- collection of user charges from the households
- Selling of compost to the local residents

Role of ULB in DSWM

- Regular collection of dry waste without fail
- Providing infrastructure like land, pushcarts etc, political, financial and technical support
- Proper collection of debris, biomedical and industrial waste preventing it from getting mixed with the municipal waste
- Identify resource persons for creating awareness and to provide sound technological guidance on composting
- Purchasing the compost for public parks, gardens of government institute and roadside plants

Role of RWAs in DSWM

- Enhance source segregation and proper door to door waste collection
- Regularly monitor the activities of SWM
- Conduct regular awareness programmes for the waste retrievers on importance of segregation.
- Encourage proper separation of recyclables by retrievers by by allowing them to keep the amount gained from selling these recyclables as incentive.
- Collect proper user charge from each household
- Try involving more number of households in this activity.

Types of Composting adoptable in the Neighbourhood Area

- *Aerobic composting*: Aerobic is the system utilizing oxygen. This method is the most simplistic and successful, even though the minimal loss of nitrogen through free elements into the atmosphere is present.
- *Anaerobic composting*: Anerobic composting requires storage

tanks and other systems to keep oxygen from invading the anerobic decay process. This method also takes longer to complete and manufactures methane gas as a by-product. The odor is disagreeable, the gas itself is flammable, and should be vented off periodically. It is generally ill suited for the home composter.

- *Vermin composting*: is a composting of organic waste with the help of Earthworms in the presence of oxygen. The only disadvantage of this process is that it requires proper care and maintenance.

Many such examples of decentralized solid waste management system has been worked out in Bangalore with the help of Bangalore Mahanagara Palike (BMP).

Scientific Handling of Waste Society (SHOWS) is one of such working examples for decentralized solid waste management system. They have developed an effective system for SWM in which 75% of the waste is recycled and only 25% goes to the landfill area thereby saving 75% of transportation and landfill operational charges for BMP.

Shows is a group of residents who volunteered in the year 1995 to establish their own SWM system with the help of Bangalore Mahanagara Palike. In this system, 75% of the waste is recycled and the rest i.e. 25% is sent to landfill area.

The process involved is explained step by step:

- The waste is collected door to door from 500 households. 300 Kgs of waste is collected every day.
- Segregating the waste on spot into four parts i.e. compostable, recyclable, refuse and garden waste.
- Proper segregation at SHOWS Park into compostable, recyclable, other plastics and non recyclable wastes.
- Composting of the compostable waste.
- Preparation of leaf mulch from the garden waste and street sweeping. Leaf mulch is a good water-retaining agent. The leaves are ratained in a separate cage for 30 days and the decomposed leaf layer collected from the botom of the cage is shifted into 3 separate stacks.
- Sale of recyclable waste and the amount gained is distributed among the workers as incentive to enhance segregation.
- Handing over the refuse to the Bangalore Mahanagara Palike trucks, this will be taken off to the dump yards.
- After 60 days, the compost is dried and aerated for 2 days and then shredded and sieved to get fine odourless compost rich in

micro nutrients. This compost is packed and sold to the local residents at the cost of Rs 5 per Kg.

- Selling recyclable waste, and delivering the non-recyclable waste to the Bangalore Mahanagara Palike trucks every morning.
- An MoU has been signed between SHOWS and BMP in 2002 for two years in which BMP supports with a non recurring expenditure of 2.6 lakhs and an annual recurring expenditure of 2.33 lakhs, including salaries of 6 employees and one supervisor. The progress of this project is regularly reviewed by BMP. SHOWS has been able to work within approved budget in the first year.
- A user charge of Rs. 300 per year is collected from each household.

General problems Faced by Shows while Establishing Compost unit

- Odour problem
- Breeding of domestic flies and their maggots
- Handling of leachate
- Extended duration of composting period

Shows has successfully overcome all these problem through an indigenous technology developed by them. It has come out with a foolproof technology for aerobic composting of the organic waste. The compostable waste is spread in the stacks, on the powder-coated grids (to reduce corrosion of mesh) placed at 6 inches height from the ground to avoid soil contamination.

Underneath the mesh, the ground is of double 'V' shape with a gradual slope towards the outlet mouth. The leachate is treated for malodour with the diluted herbal deodourising solutions, introduced from the inlet pipe.

The outlet mouth will be screwed open and the treated leachate is collected in buckets, which is used for watering a native species of Gobara trees, which absorbs lots of Nitrogen. Proper care is taken that this leachate is not drained in the sewers. The waste will be covered with a thin layer of leaf mulch, avoiding smell, reducing leachate generation and enhancing the decomposition process. In the initial period, the temperature inside the stacks rises to 60 to 70 o C killing all the pathogens present in the waste.

To quicken the decomposition process and to lessen odour, air is blown in to the stacks. The air pipes are placed in the 6-inch gap with side spores in it. The stacks are covered with a tray filled with charcoal and leaf mulch to avoid smell, flies and animals. The waste is held for

50 to 60 days in 11 such stacks and the air is blown at a regular interval. The blowers are not operated during early morning and afternoon to reduce sound affects on the residents.

SOUND ENVIRONMENTAL MANAGEMENT OF SOLID WASTE - THE LANDFILL BIOREACTOR

The Management dilemma of Solid Wastes (SW) has been recognized and tackled seriously worldwide. However, in developing countries the problem is still hindered by social and economic predicaments and priorities.

In our region, the generation of solid wastes has become an increasingly important environmental issue over the last decade, due to the escalating growth in populations and the changing life style, leading to new trends of unsustainable consumption patterns concomitant with inflation in wastes production.

Such increases in solid wastes generation concurrent with shifting characteristics pose numerous questions concerning the adequacy of conventional wastes management systems, and their associated environmental, economic and societal implications. The majority of the Arab Countries are still striving towards establishing suitable and practical mechanisms for addressing the MSW issue at local and national levels.

The root of the problem and the variability in its magnitude and implications rely solely on the specificity of inherent local conditions and the availability of financial and technological capabilities of the local governorates, townships and municipalities. Most national Arab legislations and strategies are instituted in accordance with the universally common integrated MSW Management, i.e., based on the "cradle-to-grave" principles.

Nonetheless, as sound as the approach is theoretically, it is still impeded by the lack of reliable databases on MSW generation and classification, shortage in infrastructures suitable for comprehensive waste management (considering its variable characteristics), as well as deficit in trained personnel to undertake the various aspects of the problem systematically and holistically.

Many cities and municipalities in the region are still faced with serious challenges emanating from the vast expansion in urban developments, the increasingly unsustainable life style and consumption patterns, and the absence of feeand/ or tax-based administrative structures and fiscal resources. The development of Integrated Solid Waste Management (ISWM) systems necessitate the establishment of a hierarchy based on a sound pyramid of prevention, source reduction and waste minimization, resource recovery, reuse and recycling, waste neutralization and treatment, combustion and landfilling practices.

These elements collectively provide the basis for a sound management strategy of solid wastes; nonetheless, specific conditions and capabilities must be stressed in putting more emphasis on some versus the others.

However, should the integrated management system of wastes be unachievable holistically for lack of resources, emphases on proper collection and safe disposal must be pursued as the least acceptable solution.

Table Elements of Integrated Solid Waste Management (ISWM):

- Pollution Prevention, or at source waste minimization (or reduction) in quantity, volume and/or harm.
- Resource Conservation, or environmentally sound reuse and recycling, and recovery (e.g., composting or energy recovery).
- Sound Treatment to render safe; and
- Safe Ultimate Disposal (Incineration or landfilling).

The pertinent areas that can be covered by Sound Practices (SP) of ISWM are: Sorting and Special Wastes, Waste Reduction, Collection and Transfer, Recycling, Composting, Incineration, and landfilling. Of all available disposal options under the sound solid waste management system, landfilling method is the most commonly utilized worldwide, and perhaps the most favourable (especially in developing countries). Landfills form the backbone of every integrated waste management plan.

Historically, landfills have served as ultimate waste receptors for municipal refuse, industrial discards, agricultural residues, wastewater sludge, incineration ashes, as well as hazardous wastes. Uncontrolled dumping of waste into the local environment should be replaced however, as soon as possible with controlled sanitary landfilling along with other treatment and disposal methods. The most noticeable problems associated with the landfilling practice are the generation of leachate and gas.

In any case, properly sited, constructed, operated and monitored landfills offer a safe disposal approach for disposing of municipal (and other) solid waste, typically at the lowest cost compared to other disposal options. Because of their practical merits and economic favourability landfills have promoted greater interest in their use, innovation and advancement.

All landfill sites, as an integral part of solid waste management systems, should be operated to a standard, which protects human health and the environment. There are four known types of landfills: Uncontrolled Dumps, Total Containment, Modified Containment and Controlled Release.

The first type (common in many developing countries) is not a management method. The other types are indeed different forms of the engineered "Sanitary Landfill". The second landfill type attempts per se to prevent all inflow and outflow of liquids into the landfill system, slowing down the rate of waste degradation and impeding stabilization (sealed entombment approach). However, since total containment cannot be practically guaranteed, this type is not usually recommended.

Modified containment landfill allows some surface water to penetrate and the surplus leachate to flow through to be later collected and removed. This third type is technically preferred and can be accompanied with leachate-recirculation in order to control the microbially mediated decomposition processes, accelerating wastes stabilization; such a special type is known as the "Landfill Bioreactor". Finally, controlled release landfills can be used for inert wastes such as construction debris and incineration ash.

As for the ultimate disposal of hazardous wastes, a unique type of landfilling known as the "Secure Landfill" should be employed to provide extensive protection of human health and the environment against the perceived risks. This paper will identify and assess the current status and conventional practices concerning MSW management in the Arab region.

Also, it will address the various options of solid wastes management options with a particular focus on landfills due to their importance especially in developing countries. Furthermore, it will draw comparisons, conclusions and recommendations in this regard, and point out workable solutions.

SOLID WASTE MANAGEMENT IN THE ARAB REGION

Population increase, economic progress, expansion in urban areas, rapid industrial development, and rising standards of living have all contributed to a sharp rise in solid waste generation in the West Asia Region. It is estimated that municipal waste generation rates in West Asian Arab states have increased from 4.5 million tons per year in 1970 to a staggering 25 million tons in 1995.

The annual rate of municipal solid waste increase is about 3.3 per cent, which is on average equivalent to that of population increase. These figures, however, can be underestimated indeed, since there are no reliable statistics on actual amounts of waste produced in many countries of the region. A quick prediction of the amounts of municipal solid wastes generated in the Arab Region in the Year 2020 shows that this figure can simply exceed 200 million tons per year.

Some statistical figures on municipal solid waste in the Arab Region are provided in Tables, respectively. It is of importance to indicate the obvious correlation between the per capita income and the rate of solid waste generated, i.e., higher GDP causes higher waste generation. From the Tables below, one can notice that 80% of the total municipal solid wastes in the Arab Region are decomposable and recyclable, and the remainder is inert matter. On average in the region, approximately 50 to 60% of municipal solid wastes are organic, and about 10% paper, 7% plastic, 4% glass, 4% metals, and 4% textiles.

Table: Solid Wastes Generation in Some Arab Countries

Country	Municipal waste kg/Capita/day	Country	Municipal waste kg/Capita/day
Bahrain	1.6	Qatar	1.3
Egypt	1.2	Saudi Arabia	1.3
Jordan	0.9	Syria	0.5
Kuwait	1.8	Tunis	0.6
Oman	0.7	UAE	1.2
Morocco	0.33	Yemen	0.45

Table: Solid Wastes Generation and Composition in the Arab Region

Countries	Kg/capita /Year	Organic Matter,%	Paper, %	Plastic, %	Glass, %	Metals, %
Bahrain	584	59.1	12.8	7.4	3.4	2.1
Iraq	285	63	1	1	1.6	1.1
Jordan	330	63	11	16.8	2.1	2.1
Kuwait	660	51	19	13	4.5	5
Lebanon	220	59	18	8	8	2.4
Oman	256	60	8	12	10	9
Qatar	475	57	18	12	3	5
Syria	185	62	4	7	4	6
UAE (Dubai	750	42	6	10	3	3
Abu Dhabi	542	49	6	12	9	6
Yemen	165	55	14	13	1.5	2

The integrated waste management strategy (i.e., from cradle-to-grave) with its universal hierarchy has been introduced recently in several countries as well. Modern collection, treatment and disposal systems have started to be employed in the region, such as vehicle collection and sorting, composting, incineration of medical wastes, and sanitary landfilling. Recycling, reuse and recovery are still at infancy stages, nonetheless are gaining popularity. Due to poverty, scavenging still exists in some areas, providing recycling and unintended environmental benefits among economic earnings.

In the Arabian Gulf sub-region, waste collection and disposal are highly efficient, and, sanitary landfills are widely used. The high content of organic matters in municipal waste triggered the interest in composting, thus several composting plants have been established and are already in operation producing compost -fertilizing materials and soil conditioners. The municipality manages for instance in Doha-Qatar, a large composting plant since the eighties. Some fractions of the waste, e.g., aluminum, paper, glass and plastics are collected and recycled in small recycling plants scattered in the region, that are mainly operated by the private sector.

The global practices of ISWM vary from region to region, country to country, and from one municipality to another, depending upon the prevailing specific conditions (natural, social, economic...etc.). Table provides an overview of some wastes management practices in several industrialized countries. Landfilling is overwhelmingly the most common waste management practice, because of its practicality, simplicity, effectiveness and economic favourability. However, there is an undeniable trend towards increasing reuse/recycling of wastes components around the world.

For instance in the USA, recycling activities increased from mere 6% of total municipal solid wastes generated in the 1960s to about 30% in the Year 2000. On the other hand, combustion has declined from 30% to less than 15% during the same period. As in other developing countries, there are numerous difficulties facing Arab member states in their endeavors to develop and implement ISWM. These may be summarized as economical, technical, institutional, social and attitudinal.

The situation is perhaps more satisfactory in the GCC countries due to economic affluence. Based on a field survey study conducted in several Arab cities, the current management practices pertinent to municipal solid wastes are given in Table. Note that landfilling is the predominant approach to MSWM in the Arab Region. Nonetheless, composting should also be considered seriously, along with sorting and recycling, due to the high portions of organic degradable fractions ar d recyclables materials generated from the MSW.

Table: Solid Wastes Generation and Composition in Some Arab Cities

Location / Category	Aden	Aleppo	Amman	Bahrain	Cairo	Kuwait	Riyadh	Tunis	Bombay	USA avg.
Food Wastt,%	57.1	59.4	54.5	59	67	50	3	68	58	11.2
Paper & boards,%	10.7	13.1	14	12.8	18	20.6	31	10	10	37.4
Plastics, %	10.8	11.5	13.2	7.44	3.4	12.6	2	11	11	10.7
Metals,%	5	0.8	2.4	2.05	2.2	2.6	16	4	2	7.8
Glass,%	2.7	7.6	2.8	3.39	2.5	3.3	3	N	3	5.5
Wood,%	*N	0.5	N	N	N	4.8	10	N	N	5.5
Textile, rubber & leather,%	5.6	3.7	4.7	6.92	0.5	4.8	2	2	N	6.7
Yard timmings %	N	N	N	N	N	N	N	N	N	12
Other,%	8.1	3.4	8.4	8.4	6.3	1.3	2	5	16	3.2

Table: Solid Wates Management Practices in Some Industrialized Countries

Countries	Kg/ Capita /day	Landfilling, %	Incineration, %	Composting %	Recycling %
Canada	1.65	80	6	4	10
Germany	0.95	45	35	4	16
Italy	1.1	75	13	7	5
Japan	1.26	15	60	5	20
Spain	0.95	65	5	17	13
Switzer land	1.2	10	58	10	22
UK	1.15	85	8	2	5
USA	1.98	65	10	2	23

Table: Current SWM Practices in Some Arab Cities

City	Landfilling	Incineration	Composting	Recycling
Aden	Y	N	N	P
Aleppo	Y	N	N	P

Amman	Y	P	N	P
Bahrain	Y	P	N	P
Cairo	Y	P	P	P
Kuwait	Y	P	P	P
Riyadh	Y	P	P	P
Tunis	Y	N	P	P

THE LANDFILLING OPTION

Integrated solid waste management (ISWM) is a concept that goes beyond the mere safe disposal of wastes. It addresses the root cause of wastes problem by emphasizing the notion of "Cradle to Grave" responsibility. The main pillars of this strategy are: Prevention, Reduction, Recovery, Treatment and Disposal. The strategy of ISWM management implies a hierarchy in the implementation methodology, by ranking the above-mentioned operations in accordance to their priority or favourability.

Nonetheless, often there are many difficult situations that face municipal managers in planning and directing concrete ISWM projects in cost-effective, technically sound and politically favourable manners. These constraints may eventually cause lesser adherences to the ISWM complete hierarchy and likely curtail conformity to more feasible options like landfilling practices. Landfilling is the oldest and most widely practiced method for disposing of solid waste.

Properly constructed and operated landfill sites offer a safe disposal way for municipal (and industrial) solid wastes, typically at the lowest cost compared to other disposal options. Should the financial resources be limited (as this is the case in many developing countries), it may not be necessary from health and/or environmental viewpoints to invest in other disposal methods given that suitable sites are available for landfills. In other words, sanitary landfills can be "stand alone" facilities under constrained conditions. Landfills however, form the basis of every integrated solid waste management plan.

There are a number of complementary (or alternative) treatment options for waste, e.g., incineration, composting...etc., but none of these treatment options can function alone. All require landfill as a final stage: all other waste management options, such as recycling and incineration, rely on landfills for disposing of unsuitable refuse or inert residues, respectively. Therefore availability of some landfill space is essential for every region, and will continue to be so in the future in spite of technological advancements. There are four universal types of landfills: Uncontrolled Dumps, Total Containment, Modified Containment and Controlled Release. The first type (common in many underdeveloped countries) is not at all a waste management method. The other types are indeed different forms of the engineered "Sanitary Landfill". The second landfill type attempts per se to prevent all inflow and outflow of liquids into/from the landfill system, slowing down the rate of waste degradation and impeding stabilization (sealed entombment approach).

However, since total containment cannot be practically guaranteed, this type is not usually recommended. However, some of its applications can be utilized for hazardous waste disposal with repetitive layers of containments. Modified containment landfill allows some surface water to penetrate and the surplus leachate to flow through to be later collected and removed. This third type is technically preferred and can be accompanied with leachate-recirculation in order to control the microbially mediated decomposition processes, accelerating wastes stabilization; such a special type is further developed to a new family known as the "Landfill Bioreactor".

Finally, controlled release landfills can be used for inert wastes such as construction debris and incineration ash. As for the ultimate disposal of hazardous wastes, a unique type of landfilling known as the "Secure Landfill" should be employed to provide extensive protection of human health and the environment against the perceived risks. The criteria for selecting landfill sites and the operational methods employed have been developed scientifically with cognizance of the intrinsic stabilization processes occurring within landfill as well as with due considerations to the environmental and health impacts associated with landfilling.

The most perceptible problems associated with landfilling practice are the generation of leachate and gas. Leachate is mainly produced by infiltration of rainwater into the solid wastes, followed by the potential movement of leachate into an under-drain collection system or into the surrounding environment if containment is inadequate. The leachate may contain high concentrations of organic and inorganic materials, including toxic compounds and heavy metals. Another potential environmental risk is also linked with the production and migration of gases from the waste during landfilling and upon decomposition.

Landfill gases (e.g., CO_2, CH_4, H_2S, and Volatile Organic Compounds (VOCs)) are mainly resultants of the anaerobic decay of solid wastes within the site. Methane usually accounts for 60% of the total gas production in a landfill; the gas is flammable and can be explosive when present with sufficient amount of oxygen, and is also implicated as a greenhouse gas. Carbon dioxide can produce carbonic acids, lower pH and cause mineralization in groundwater. Both of these gasses are considered Greenhouse Gasses (GHG). VOCs, usually toxic and/or carcinogen, often depend in their characteristics and contents on the properties of the solid wastes disposed (e.g., amount and type of wastes and hazardous components and/or mixtures).

The dissolution of heavy metals can be a serious pollution problem, which may threaten groundwater resources through leachate migration. Odor problems are often caused by hydrogen sulfides production, which can also react with heavy metals, form precipitates and clog under-drain systems.

THE LANDFILL BIOREACTOR

Most sanitary landfills are traditionally constructed and operated to allow single-pass leaching of infiltrated rainfall through the refuse mass. Due to this gradual leaching process, stabilization in such landfills may require many years, thereby extending the acid formation and methane fermentation phases of waste stabilization over long periods of time. Under these circumstances, decomposition of biodegradable fractions of solid waste will be impeded and less complete, often disallowing commercial recovery of methane gas and delaying closure and possible future reuse of the landfill.

On the contrary, leachate recycling may be used as a management alternative which requires the containment, collection and recirculation of leachate back through the landfill media. This option offers more rapid development of active anaerobic microbial populations and increases reaction rates (and predictability) as well as a compressed time required for stabilization of the readily available organic constituents as low as two to three years (rather than the usual 15 to 20-year period referred to previously).

This accelerated stabilization is enhanced by the routine and uniform exposure of microorganisms to constituents in the leachate, thereby providing the necessary contact time, nutrients and substrates for efficient conversion and degradation. Hence, leachate recirculation essentially converts the landfill into a dynamic anaerobic bioreactor, which accelerates the conversion of organic materials to intermediates and end products. The landfill bioreactor can provide higher and more efficient attenuation of toxic substances, and a much lower risk of adverse public health and environmental impacts.

The enhanced attenuation capacity of such landfills with leachate recirculation is not only applicable to easily degradable and non-hazardous material, but also to toxic heavy metals and resistant organic wastes. Mobilized metals (e.g., iron, cadmium, nickel, zinc and lead), soluble during low pH conditions, can be precipitated as metal hydroxides, carbonates and sulfides, reducing the leachate concentrations to undetectable levels. Due to the accelerated and intensified reduction conditions, onset by leachate recirculation, the potential toxic effects of such metals are alleviated within the landfill.

The microbial acclimation opportunity present at landfill bioreactors due to extended residence time (warranted by recirculation), along with the enhanced contact chances will allow for the degradation of more recalcitrant and less biodegradable organic compounds. Reductive dehalogenation is a prime example of this in situ bioremediation capacity. These attenuation mechanisms are not available (or much less effective) in conventional, single-pass leaching landfills, where the "washed-out" leachate has to be routinely withdrawn and treated externally for pollutants elimination.

In addition, since methanogenesis is more profound and proceeds in shorter periods of time at leachate recycling landfills, predictable and economical recovery of methane gas can be achieved. The post-closure care is evidently more focused, cheaper and accelerated, and landfill space can be reclaimed for further use much sooner than under single pass operation. The sequential changes occurring throughout landfill stabilization processes are essentially reflected in the quantity and characteristics of generated leachate (and gas). Accordingly, an effective recirculation strategy for leachate must not only accommodate the stabilization stage of a landfill, but should also be compatible with the characteristics of the recycled leachate.

Such varying leachate characteristics, however, result from various organic transformations as well as inorganic complexation, precipitation and/or sorption. The most important phases of landfill stabilization established during leachate recycle are acid formation and methane fermentation. The former phase gives rise to high-strength and aggressive leachate of low pH accompanied by low biogas production, and the latter is categorized by lowerstrength leachate at neutral pH but with high gas production.

Accordingly, parameters of operational significance during the earlier acidogenesis phase are organic constituents (COD and VOA), pH and conductivity, and gas production and composition. Frequencies and rates of leachate recirculation during this phase should be kept low until gas production increases so that acidic inhibition of methane fermentation can be avoided. As gas production increases, the leachate loading frequency can also be increased until a routine schedule can be established. The parameters indicative of operational characteristics during methane fermentation include COD, VOA, pH, alkalinity, and gas production and composition (CO_2 and CH_4). Throughout these two phases of landfill stabilization, a combination of a weekly sampling for COD, pH and gas, monthly sampling for VOA and nutrients (N and P), and a possible special quarterly sampling for priority pollutants (hazardous organic compounds and heavy metals) my be recommended for performance monitoring, and adjustment of leachate recirculation rates and scenarios.

As leachate strength and gas production diminish with time, indicating removal and stabilization of the readily leachable and decomposable refuse factions (final maturation phase), leachate recycle should be discontinued and the accumulated pool of leachate should be removed for new applications and/or ultimate disposal. It may be feasible to supply leachates from old landfill sections into newly developed landfilling areas to initiate the recirculation programme and provide seeds for stabilization.

Prior to leachate ultimate disposal, and in addition to the routine analyses proposed previously, particular attention would need to be given to any post-treatment requirements, such as for excessive nutrients, dissolved solids, residual metals or salts, and reducing conditions (ORP). In situ treated, stabilized leachate may in certain cases be used for irrigating the vegetative covers on the site, or even evaporated. Modern landfill bioreactor with leachate recirculation may also have discrete cells to induce nitrification and denitrification.

Innovations on landfill bioreactor were also developed to integrate specialized sections of aerobic, facultative, and anaerobic treatment processes into the configuration and operation of leachate recycling landfills. Application of leachate recirculation to full-scale landfills has occurred with increasing frequency in recent years. Leachate is returned to the landfill using a variety of techniques including wetting of waste as it is placed, spraying of leachate over the landfill surface, and injection of leachate into vertical wells or horizontal trenches installed within the landfill.

It is important to design and operate other landfill components such as active gas management system; effective leachate collection system, final and intermediate covers, and site monitoring system so that they are compatible with bioreactor operations). Figure provides an illustrative configuration and engineering components of a Landfill Bioreactor.

By providing containment by liners and covers, sorting and discrete disposal of waste materials into dedicated cells, collection and in situ recirculation of leachate, and management of gas; the landfill can be essentially transformed into an engineered reactor system. Such landfills are capable of accelerated biochemical conversion of wastes and effective treatment of leachate. Accordingly, fundamental knowledge of biological degradation processes, and comprehension of landfill stabilization sequential phases are necessary for the prospective development and application of in situ treatment of leachate.

Linking the basic understanding of the inherent, microbially mediated and physical/chemical reactions with the spatial and temporal domains of landfill will warrant successful operation of the landfill as a bioreactor system.

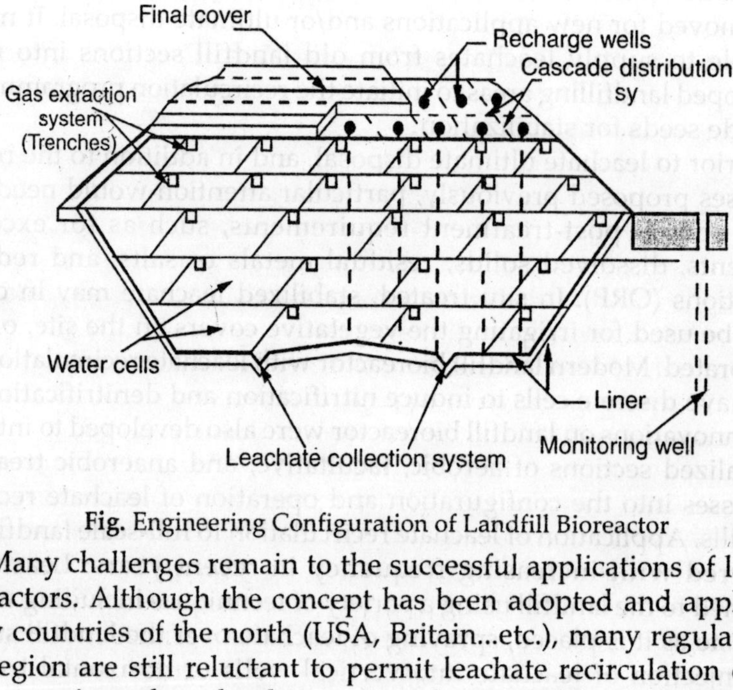

Fig. Engineering Configuration of Landfill Bioreactor

Many challenges remain to the successful applications of landfill bioreactors. Although the concept has been adopted and applied in many countries of the north (USA, Britain...etc.), many regulators in our region are still reluctant to permit leachate recirculation due to misconceptions about leachate accumulation within the landfill. The ability to uniformly wet the landfill is often limited by operational practices such as the use of impermeable daily cover as well as the inherent heterogeneity of waste.

Furthermore, landfills are too frequently designed with insufficient storage external to the leachate recirculation landfill, forcing the use of excessive recirculation rates (even during acidogenic conditions). This may not be the case in our arid dry region; however, the strength of leachate might be more potent. Co-disposal applications of municipal and industrial solid wastes, and/or with wastewater sludges should be investigated carefully in each case. Finally, landfill operators must be trained to optimize the landfill as a biological reactor rather than a long-term repository pit of waste matters.

Proper integrated management of Solid Waste (ISWM) in the Arab region is generally hampered by technical, administrative and financial shortcomings. In order to tackle this growing problem of solid wastes, systematic approaches on the local, national and regional levels should be explored and implemented, based on the prevailing conditions and priorities. Should the ISWM implementation become unfeasible holistically in certain municipalities, elements of which are to be executed to tackle the solid wastes dilemma gradually. This approach undoubtedly favours sanitary landfilling as a "standalone" option for waste disposal practices.

Landfill is an integral part and ultimate step of sound solid waste management strategy. Municipalities must make provision for future landfill needs by allocating suitable land in their long-term plans. It is crucial that each district have an appropriate site for waste disposal otherwise wastes will be illegally dumped or exported to other areas. All landfill sites should be operated to a standard, which protects human health and the environment. Where landfill standards have not reached the minimum for protection of health, it is recommended that action be taken immediately.

Continuous and thorough monitoring of waste inputs, gaseous emissions and liquid effluents is essential for successful landfilling operations and protection of public health and the environment. With proper engineering design and careful operations, landfills using leachate recirculation can be essentially converted to dynamic earthen bioreactors capable of enhancing the decomposition of solid wastes, while providing in situ containment and treatment of leachate.

Furthermore, leachate recirculation landfills or Landfill Bioreactors are superior to the conventional single-pass leaching landfills; they lower the risk of adverse impacts of waste disposal by providing better engineering control and attenuation capacity against toxic substances.

Chapter 7

Radiation in Our Environment

The purpose of this factsheet is to provide some basic information about radiation and radioactivity to inform our discussions about radioactive contamination in the environment and its possible effects on our health.

RADIATION AND RADIOACTIVITY

Radiation is energy that travels in the form of waves or particles. There are many different kinds of radiation—such as microwaves, visible light, and X-rays – with a wide range of energies . Radioactivity is the property of some atoms to spontaneously give off high-energy radiation in the form of particles or rays. The atoms that make up radioactive materials are the sources of radiation and are called radionuclides.

IONIZING AND NON-IONIZING RADIATION

There are many ways to categorize the different types of radiation. In this factsheet we will refer to non-ionizing radiation and ionizing radiation. Non-ionizing radiation has enough energy to move atoms in a molecule or to cause them to vibrate, but it does not have enough energy to change them chemically. Visible light and microwaves are examples of non-ionizing radiation.

When an unstable atom decays, it can give off enough energy to affect the structure of other atoms and to break chemical bonds. This energy, called ionizing radiation, travels in the form of subatomic particles or waves. Ionizing radiation is commonly used to generate electric power, to kill cancer cells, and in many manufacturing processes. When people think of radiation, most often they are thinking of ionizing radiation.

TYPES OF IONIZING RADIATION

There are four main kinds of ionizing radiation . Alpha particles have two protons and two neutrons; alpha particles tend to be slow moving and can usually be stopped by a piece of paper or the dead outer layer of skin tissue. Beta particles are essentially electrons; they are smaller and faster moving than alpha particles and can travel significant distances in air but will usually be stopped or reduced by clothing or a few millimeters of glass or aluminum.

Gamma rays are packets of energy (photons) without mass or charge; most gamma rays can pass through or be absorbed by the human body but will lose most or all of their energy in a few feet of concrete or soil or several inches of lead. X-rays are like gamma rays but have less energy and are less penetrating; they can usually be stopped by a few millimeters of lead.

RADIATION IN OUR ENVIRONMENT

Radiation is all around us. Some radioactive materials occur naturally, and others are created by humans. Human exposure to naturallyoccurring radiation can also be increased by human activities such as mining or air travel. Many naturally occurring materials in our environment are radioactive.

For example, thorium and potassium in the earth's crust are naturally radioactive, and radon is a naturally occurring gas in most soils. Humans create and use radiation sources in a variety of applications, such as electricity generation, healthcare, scientific research, and food irradiation.

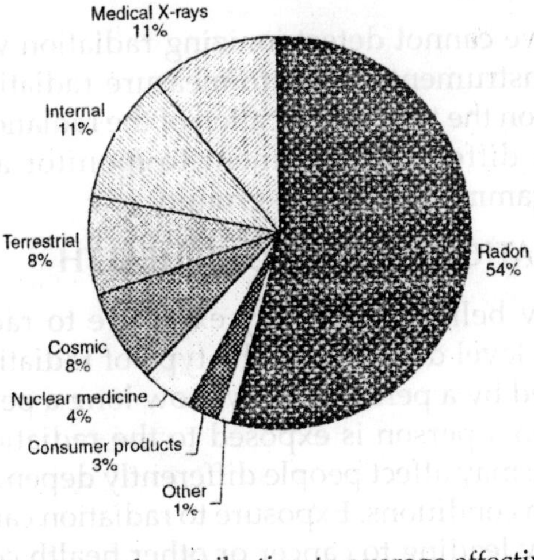

Fig. Annual contribution to average effective radiation exposure for people

When we talk about human exposure to radiation, we often think of nuclear weapons. Detonation of nuclear weapons exposed people nearby to high doses of radiation, and global background radiation levels increased significantly in the short-term after detonations. Fallout from atmospheric detonation of nuclear weapons peaked in the mid-1960s. The longterm increase in global radiation levels from nuclear bomb detonations, while measurable, is considered to be statistically small.

Radionuclides Move though Our Environment and into Our Bodies

Radionuclides can be found in air, water, soil, and rock. They may occur there naturally or may have been released as a result of human activity. Radionuclides can also enter the human food chain through plants and animals.

Not all radionuclides move through the environment in the same way. For example, one radionuclide may dissolve in water while another might tend to stick to the surface of soil through which water is moving.

Radionuclides may get into our bodies through the skin or an open wound, or when we swallow or inhale them. What happens to the radionuclide once inside the body depends on the specific radionuclide. For example, some may stay in the body for a long time while others might be exhaled or eliminated quickly.

Detection of Ionizing Radiation

As humans we cannot detect ionizing radiation with our senses: we have to use instruments to try to measure radiation. The type of detector depends on the type and location of the radiation. For example, you might use a different instrument to monitor air-borne alpha particles versus gamma rays in the ground.

IONIZING RADIATION AFFECT YOUR HEALTH

It is currently believed that any exposure to radiation may be harmful. The risk level depends on the type of radiation, the amount of energy absorbed by a person's body, how long a person is exposed for, and how often a person is exposed to the radiation. In addition, the same exposure may affect people differently depending on age and pre-existing health conditions. Exposure to radiation can cause damage to cells, eventually leading to cancer or other health concerns.

Exposure to radiation can also cause damage to a person's genes, potentially causing problems for any offspring. While exposure to high doses of radiation may result in illness and/or death in a few weeks to months, health problems from low or moderate exposure may not show up for months or years. It is believed, however, that the health risks from low-level exposure are very small.

ADAPTIVE RADIATION IN A FLUCTUATING ENVIRONMENT

The emergence of diversity within a rapidly growing population is a major evolutionary feature known as adaptive radiation. It occurs when a single ancestor diversifies to use a range of environments where each type differs in traits that are used to exploit those environments. The principal explanations for this widespread phenomenon focus either on divergent selection generated by environmental heterogeneity both in space and time, or divergent selection generated by competition for limiting resources that drives closely related species to exploit different parts of a resource spectrum. Great progress has been made in our empirical and theoretical understanding of the workings of these processes.

However, much of our knowledge of adaptive radiation is limited to either controlled laboratory experiments in constant environments in which selection is considered to be continuous through time, or to studies of natural populations where the environment cannot easily be controlled. As a result, we know little about the effects of disturbance on adaptive radiation and in particular whether intermittent disturbance hinders or favours the process. Given the ubiquity of environmental disturbance across all scales of space and time, this is a surprising observation. To address this issue we studied the effects of disturbance upon the evolution of diversity in a microbial microcosm.

Following Gerritsen and Patten, we define disturbance rather generally as an 'infrequent event or episode that causes a system to deviate from its expected or regular state'; this definition subsumes many of the previous uses of the term in the literature. The relationship between disturbance and the maintenance of diversity is a classical question in community ecology that has received considerable theoretical and empirical attention. However, the effect of disturbance on the evolution of diversity has rarely been considered.

To ascertain how disturbance causes a diversifying population to deviate from its expected trajectory, it is necessary to consider the type of disturbance and how it may influence the rate of diversification and the steady-state level of diversity. We may generate several hypotheses for these effects:

- By intermittently redistributing and even renewing resources, disturbance may limit the intensity of divergent selection for the limiting resource. Also, by intermittently redistributing individuals across the landscape, disturbance would hinder evolutionary divergence due to the aggregation of different morphological types across a heterogeneous environment. Both these effects would act to delay the onset of divergence. The first effect would reduce divergent selection; the second would reduce the evolutionary response to divergent selection.

- If disturbance (e.g. variable temperature) slows the rate of resource consumption, then the rate of diversification due to resource competition will also be slowed.
- By imposing mortality, disturbance may eliminate rare species; if the rate of extinction in a disturbed environment is greater than that in an undisturbed environment, then the steady-state level of diversity will be lower in the former than the latter.
- Disturbance may constrain the number of types that can exist at steady state by determining the frequency of certain habitats that would otherwise exist in the absence of disturbance. This is just a reformulation of the intermediate disturbance hypothesis cast at an evolutionary time scale that predicts a non-monotonic relationship between disturbance frequency and local diversity.

The effect of disturbance will also depend upon its temporal structure. Different disturbance types (fire, storms, etc.) can have different magnitudes, frequencies and temporal correlation. One general property of environmental variation of particular importance for ecological and evolutionary dynamics is its autocorrelation structure, which defines the statistical dependency of successive events in a series. In contrast to random fluctuations where successive values are statistically independent, autocorrelated fluctuations have serial dependency.

This property of environmental variability has been the focus of much recent study because the fluctuations of many physical variables – both 'normal' (such as temperature and rainfall) and 'catastrophic' (climate shifts, earthquakes, hurricanes) events—are positively autocorrelated across a range of scales. For example, Caswell and Cohen suggested that for a given frequency of disturbance, the temporal correlation of a series of disturbances may alter the outcome of interspecific competition, making co-existence more difficult (diversity was lower) in autocorrelated environments.

Since then, several theoretical studies have demonstrated that extinction probabilities and various aspects of population dynamics are sensitive to the autocorrelation structure of the environment. To our knowledge, however, this aspect of the temporal structure of disturbance has not been treated within the context of the evolution of diversity. Our model of adaptive radiation is a rapidly growing and evolving microbial population of *Pseudomonas fluorescens*.

This bacterium shows rapid and repeatable adaptive radiation in vitro and is thus an ideal model system for our purposes. Previous work has demonstrated the evolution of a phenotypic polymorphism from a monomorphic ancestor to be linked to the presence of spatial structure that develops in an unmixed liquid habitat. When polymorphic populations of P. fluorescens are switched to a well-mixed, unstructured habitat, the polymorphism is lost. It is thought that in the highly nutrient-rich medium rapid growth of the isomorphic population of this strictly aerobic bacterium generates strong intraspecific competition for oxygen.

Competition creates divergent selection for different phenotypes adapted to different spatial locations in the habitat: planktonic versus surface- and bottom-dwelling films. In this study, we ask whether random and autocorrelated disturbance can alter the dynamics of diversification in P. fluorescens. We studied the effects of disturbance by abruptly homogenizing the habitat. This had the effect of redistributing individuals and the limiting resource (oxygen) without directly inducing mortality.

We observed that disturbance significantly delayed diversification and that this delay was higher when disturbance was autocorrelated. Although the rate of accumulation of diversity was slower in the disturbed environments by the end of the experiment, diversity had converged to a similar steady-state level across disturbance treatments.

MATERIALS AND METHODS

Biological Material

We used strain SBW25 of the Gram-negative bacterium Pseudomonas fluorescens because it exhibits rapid and repeatable in vitro phenotypic (characterized by colony morphology when plated on agar) and ecological (characterized by the place colonized in the culture medium) diversification from an isomorphic population. Diversification occurs strictly in static liquid conditions and is inhibited – the culture remains isomorphic – in a shaken medium.

Three categories of morphotypes that arise over time were originally defined by Rainey and Travisano: SM (Smooth Morph), the ancestral morphotype, is planktonic and occupies all parts of the culture medium; WS (Wrinkly Spreader) forms a film at the surface of the medium; FS (Fuzzy Spreader) occupies the bottom of the culture medium, where it also aggregates to form a film. However, as in previously published studies with SBW25, we identified four additional types within the WS form that show different morphological features and can be readily distinguished by independent plate readers. Our diversity counts thus included a total pool of seven morphological types.

Experimental Design

We used a randomized design with destructive sampling of replicates through time. One hundred and eighty 15-ml sterile plastic tubes containing 6 ml of King's medium B (KB) were inoculated with an isomorphic (SM) population of P. fluorescens strain SBW25 at a density of approximately 106 CFU·ml^{-1}. These microcosms were kept at 28°C in a temperature-controlled incubator.

Three treatments, each with five replicates, were used:

- No disturbance,
- Autocorrelated disturbance
- Uncorrelated disturbance .

A disturbance when it occurred (the minimum inter-disturbance duration was 3 h or approximately every two generations) consisted of 2 min of orbital rotation at 25 rev·min^{-1} in an orbital culture rack (Fisher Scientific). We destructively harvested five microcosms per treatment on 12 occasions over the 7 days of the experiment. The sampling events were distributed as follows: every 8 h over the first 2 days, every 12 h over the third day, and then every 24 h for the remaining 4 days of the experiment. On each sampling occasion, all replicates were homogenized by vortex. Morphological richness and density were estimated by colony counts after dilution in KB, plating on KB agar and culture at 28°C for 48 h. For each replicate, at least two Petri dishes with 40–400 colonies were counted. To ascertain the repeatability of the results, the experiment was then repeated in full changing only the disturbance series.

Generation of Disturbance Series

We used a first-order autoregressive model to generate time-series of environmental disturbance. A homogeneous Markov chain model of dichotomous environmental disturbance yields a variance spectrum that can be adjusted from strongly autocorrelated to uncorrelated. The transition between the two environmental states (disturbed or undisturbed) was governed by the following transition matrix:

$$T = \begin{bmatrix} 1-\lambda & \lambda \\ \lambda & 1-\lambda \end{bmatrix}$$

where $0 < \lambda \leq 0.5$ and λ is the probability of a change in state; we therefore assumed a stationary environmental process. When $0 < \lambda < 0.5$ the disturbances are autocorrelated, and when $\lambda = 0.5$ the fluctuations are uncorrelated.

Two replicate series of autocorrelated ($\lambda = 0.1$) and uncorrelated disturbances ($\lambda = 0.5$) were generated within MATLAB (version 5.0), one set for each experiment. Because the number of disturbances (frequency) was held constant for both disturbance types across both experiments, all other statistical properties (mean and variance), besides the pattern of correlation, were held constant.

STATISTICAL ANALYSES

Analyses were conducted with SAS-STAT for Windows (version 8.2). The effects of treatment, time and the interaction between treatment and time on the number of morphs and on total population density (log-transformed) was assessed by analysis of variance

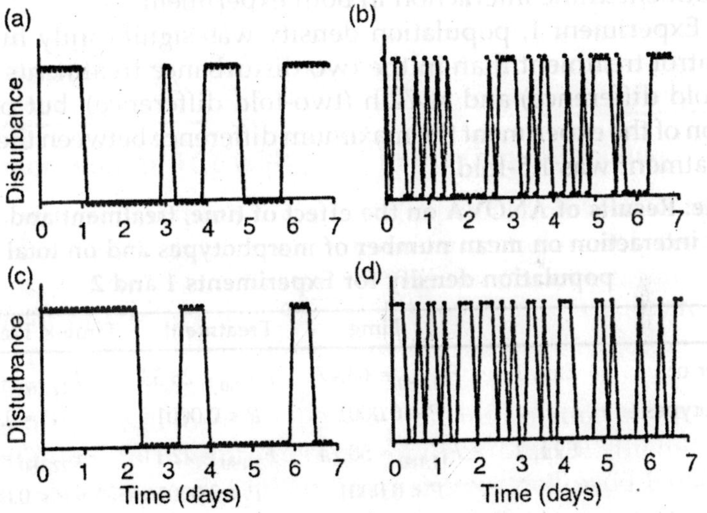

Fig. Disturbance series used in the two experiments. Experiment 1: (a) autocorrelated series ($\lambda = 0.1$), (b) uncorrelated series ($\lambda = 0.5$). Experiment 2: (c) autocorrelated series ($\lambda = 0.1$) and (d) uncorrelated series ($\lambda = 0.5$).

(ANOVA). Destructive sampling at each time period ensured that each replicate was statistically independent so that repeated-measures ANOVA was not required.

Divergence time was measured as the earliest time step showing an average number of morphs greater than 1, for each treatment in both experiments, and was assessed with a one-tailed t-test.

DIVERSITY

n both experiments, there were significant effects of the disturbance treatment, time and the disturbance *x* time interaction on the number of morphotypes.

Interaction between time and disturbance treatment can be seen in the dynamics of morphotype richness in Figure. Early in the experiment, mean morphotype richness in the disturbance treatments was lower than the undisturbed control (from a two-fold to a five-fold difference depending on experiment and time); however, this difference diminished over time as richness converged to the steady-state level in the control.

TOTAL POPULATION DENSITY

Again differences in the dynamics of the average total population density over time resulted in significant effects of treatment, time and the treatment *x*time interaction in both experiment.

In Experiment 1, population density was significantly higher in the control treatment than in the two disturbance treatments at 48 h (four-fold difference) and at 72 h (two-fold difference), but over the duration of the experiment the maximum difference between the means per treatment was 1.5-fold.

Table. Results of ANOVA on the effect of time, treatment and their interaction on mean number of morphotypes and on total population density for Experiments 1 and 2

		Time	Treatment	Time × Treatment
Number of Morphotypes	Exp. 1	$F_{11,180} = 68.47$ $P < 0.0001$	$F_{2,180} = 28.53$ $P < 0.0001$	$F_{22,180} = 1.97$ $P = 0.01$
	Exp. 2	$F_{11,180} = 50.74$ $P < 0.0001$	$F_{2,180} = 72.11$ $P < 0.0001$	$F_{22,180} = 3.37$ $P < 0.0001$
Total Population Density (ln)	Exp. 1	$F_{11,180} = 35.25$ $P > 0.0001$	$F_{2,180} = 5.04$ $P = 0.01$	$F_{22,180} = 1.73$ $P = 0.03$
	Exp. 2	$F_{11,180} = 93.01$ $P < 0.0001$	$F_{2,180} = 11.21$ $P < 0.0001$	$F_{22,180} = 3.79$ $P < 0.0001$

However, in Experiment 2, it was only in the autocorrelated disturbance treatment, and at the beginning of the experiment (up to 40 h), that the total population size was significantly lower (3.5-fold difference at 32 h) than the other treatments.

Time series of the evolution of morphotype richness over the course of the two experiments: (a) Experiment 1, (b) Experiment 2,

Diamonds and continuous line: control treatment; circles and broken line:autocorrelated treatment; squares and dotted line: uncorrelated series. Error bars 1 standard deviation.

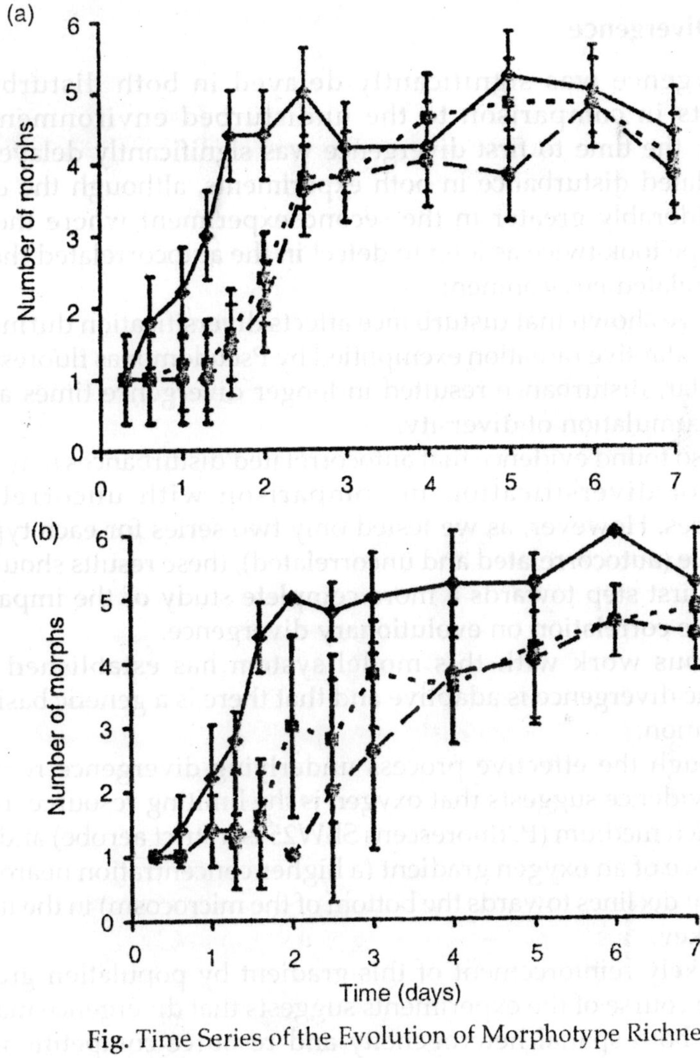

Fig. Time Series of the Evolution of Morphotype Richness.

MORPHOTYPE DYNAMICS

The disturbance treatments impaired both the formation and growth of the WS and FS morphs, which appeared later and reached lower densities than in the undisturbed treatment. More specifically, WS morphs were first observed 16 h on average after the experiment started in the control treatment, whereas WS morphs appeared after 24 h on average under the uncorrelated disturbance regime, and after

24 and 40 h on average in Experiment 1 and 2 respectively under the autocorrelated disturbance regime. In Experiment 2, FS never reached detectable densities in the autocorrelated disturbance regime.

Time to Divergence

Divergence was significantly delayed in both disturbance treatments in comparison to the undisturbed environment. In particular, the time to first divergence was significantly delayed by autocorrelated disturbance in both experiments, although the delay was considerably greater in the second experiment where the WS morphotype took twice as long to detect in the autocorrelated than in the uncorrelated environment.

We have shown that disturbance affects diversification during the process of adaptive radiation exemplified by Pseudomonas fluorescens. In particular, disturbance resulted in longer divergence times and a slower accumulation of diversity.

We also found evidence that autocorrelated disturbances may slow the rate of diversification in comparison with uncorrelated disturbances. However, as we tested only two series for each type of disturbance (autocorrelated and uncorrelated), these results should be seen as a first step towards a more complete study of the impact of disturbance correlation on evolutionary divergence.

Previous work with this model system has established that phenotypic divergence is adaptive and that there is a genetic basis for diversification.

Although the effective process underlying divergence remains unclear, evidence suggests that oxygen is the limiting resource in this nutrient-rich medium (P. fluorescens SBW25 is a strict aerobe) and that the existence of an oxygen gradient (a higher concentration nearer the surface that declines towards the bottom of the microcosm) in the liquid habitat is key.

The likely reinforcement of this gradient by population growth during the course of the experiments suggests that divergence may be driven by both spatial heterogeneity and resource competition, the latter reinforcing the pre-existing oxygen gradient in the liquid habitat. This apparently simple evolutionary system therefore belies a surprisingly complex ecological interaction.

By rapidly homogenizing the medium, the disturbance we imposed redistributed individuals and oxygen (resource) within the habitat; an effect that destroyed the habitat structure but did not result in mass mortality, as confirmed by the small changes in population density across treatments .

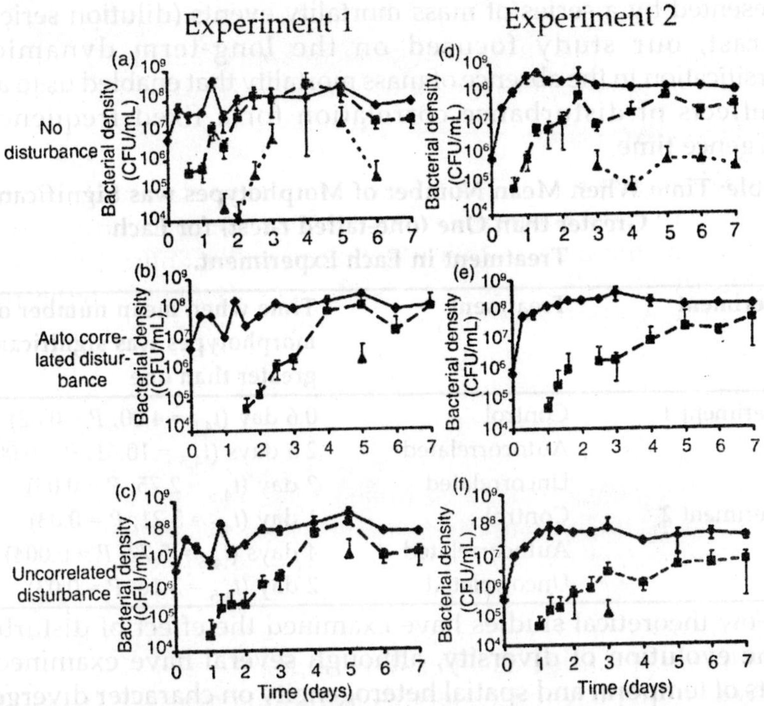

Fig. Population Densities for Each of the Three Main Morphotypes.

Population densities for each of the three main morphotypes. for each treatment over the course of the two experiments. Experiment 1: (a) control treatment, (b) autocorrelated disturbance, (c) uncorrelated disturbance. Experiment 2: (d) control treatment, (e) autocorrelated disturbance, (f) uncorrelated disturbance.

Diamonds and continuous line: Smooth Morph (SM); squares and broken line: all Wrinkly Spreader morphs (WS); triangles and dotted line: Fuzzy Spreader (FS). Error bars = 1 standard deviation. The absence of a morph means either that it was not present in the population or that its concentration was too low to be detected by plating (in this case below 104 CFU·ml^{-1}).

The disturbance thus acted directly on the principal factors—resource heterogeneity and spatial aggregation—implicated in the process of ecological divergence. We suggest that the disturbance reduced divergent selection generated by competition for oxygen and by redistributing individuals throughout the habitat may have constrained the evolutionary response to this divergent selection.

Previous experiments that focused on the maintenance of diversity (in particular the intermediate disturbance hypothesis) in the same bacterial system imposed disturbances of varying frequency represented by a series of mass mortality events (dilution series). In contrast, our study focused on the long-term dynamics of diversification in the absence of mass mortality that enabled us to assess the effects of disturbance correlation for a fixed frequency on divergence time.

Table: Time When Mean Number of Morphotypes was Significantly Greater than One (one-tailed *t*-test) for Each Treatment in Each Experiment.

Experiment	Treatment	Time when mean number of morphotypes was significantly greater than one
Experiment 1	Control	0.6 day ($t_{4,5}$ = 4.00; P = 0.02)
	Autocorrelated	2.5 days ($t_{4,5}$ = 10.61; P = 0.004)
	Uncorrelated	2 day ($t_{4,5}$ = 2.75; P = 0.05)
Experiment 2	Control	1 day ($t_{4,5}$ = 3.21; P = 0.03)
	Autocorrelated	4 days ($t_{4,5}$ = 5.72; P = 0.004)
	Uncorrelated	2 day ($t_{4,5}$ = 3.16; P = 0.03)

Few theoretical studies have examined the effect of disturbance on the evolution of diversity, although several have examined the effects of temporal and spatial heterogeneity on character divergence. For example, Day studied the impact of spatial heterogeneity on divergence and showed that features of disruptive selection depended on the movement rate between resource patches. In particular, spatial resource heterogeneity always has a disruptive effect when the migration rate between patches is low.

If we consider oxygen to be the main resource that bacteria compete for, we can consider our microcosms to be spatially heterogeneous and divided into three main regions: the air–broth interface, the bottom and the rest of the culture medium. Obviously, movement within the vial was higher when disturbance (mixing) occurred than when the microcosms remained static. As we observed more evolutionary divergence in an undisturbed environment, these results qualitatively match the predictions of this model. We also examined the effect of disturbance autocorrelation for a fixed disturbance frequency.

We chose to study the effects of temporal autocorrelation on the dynamics of diversification because it has been neglected to date both theoretically and empirically. Given that disturbance events—both large and small—are autocorrelated across a range of temporal scales, this is a surprising observation and one that may be of importance for a general theory of adaptive radiation. Although we only used two autocorrelated series and one scale of autocorrelation in these experiments, nonetheless we observed substantial and statistically significant effects on the timing of the first detectable divergence event. The only relevant study to our knowledge is that of Caswell and Cohen, who explored theoretically the impact of disturbance autocorrelation on two-species competition in a metapopulation.

They observed that in an autocorrelated environment, the time of recurrence of a species that was competitively excluded from a patch was greater than in an uncorrelated environment, which corroborates our findings. But clearly the structure and assumptions of Caswell and Cohen's model of competition in ecological time are not appropriate for this experimental model system and further theoretical work is required to establish the generality of their results within the context of the evolutionary dynamics of diversification.

RADIATION, ECOLOGY AND THE INVALID LNT MODEL

Metabolic and energetic efficiency, and hence fitness of organisms to survive, should be maximal in their habitats. This tenet of evolutionary biology invalidates the linear-nothreshold (LNT) model for the risk consequences of environmental agents. Hormesis in response to selection for maximum metabolic and energetic efficiency, or minimum metabolic imbalance, to adapt to a stressed world dominated by oxidative stress should therefore be universal.

Radiation hormetic zones extending substantially beyond common background levels, can be explained by metabolic interactions among multiple abiotic stresses. Demographic and experimental data are mainly in accord with this expectation. Therefore, non-linearity becomes the primary model for assessing risks from low-dose ionizing radiation. This is the evolutionary imperative upon which risk assessment for radiation should be based.

BACKGROUND RADIATION VARIATIONS

Exposure to ionizing radiation has always been part of the environment on Earth. Life apparently evolved spontaneously from basic chemicals formed by reactions largely initiated by ionizing radiation from the sun. When life commenced on Earth around three and a half billion years ago, the natural level of radiation was up to five times higher than today.

Everyone is exposed to background ionizing radiation in the environment from outer space, the sun, terrestrial rocks, the soil, buildings, the air that we breathe, the food we eat, and from human and animal bodies. Using a standard measure of the effects of radiation in terms of milliSieverts (mSv) per year, these exposures from natural sources total around 2.15 mSv each year at around sea level. Smaller components from man-made sources principally from medical procedures (around 0.30 mSv per year), and a minor component (around 0.05 mSv per year) from other sources including consumer products and fall-out, bring the total to around 2.5mSv per year. Doses from man-made sources therefore amount to substantially less than 20% of natural exposures. In summary, the average dose rates due to natural and man-made background ionizing radiation in industrial countries in mSv per year are:

Background radiations of natural origin	
Naturally occurring radioelements in the soil, rocks etc.	0.40
Radon, thoron, and their short lived decay products	1.20
Cosmic radiation at sea level	0.25
Radionuclides inside the human body, mainly potassium-40	0.30
Man-made sources	
Medical procedures, mainly X-rays	0.30
Other man-made sources (consumer products, air travel, fall-out etc.)	0.05

Background radiation of natural origin can be substantially higher than those around sea level. The two major sources of variation upwards are height above sea level and diverse geological backgrounds. Background radiation can shift upwards by a factor of three or more in populations with increasing altitude as found in the Rocky Mountains regions of USA. Populations numbering several millions are exposed to background radiation in the range of 3 to 8 mSv per year and much higher for some smaller groups, when exposed to geological structures rich in uranium, thorium and their decay products.

Some areas of the world, called high background radiation areas (HBRAs), have extremely high levels of background radiation. A field of medical geology has emerged defined by the International Commission of Geological Sciences for Environment Planning as the science dealing with the influence of ordinary environmental factors on the geographical distribution of health problems in humans and animals.

THE RADIATION-EXPOSURE PARADOX

Although more critical data are needed, the apparent health-promoting consequences of HBRAs are paradoxical and are frequently disputed. Definitive studies of outlier populations such as HBRAs require enormous sample sizes to differentiate between populations exposed to differing radiation levels. Even when negative correlations occur between dose and cancer incidence in these demographic studies, there is always the problem of whether correlation implies causation.

For example, Cohen found a highly significant negative association between mean radon exposure and lung mortality across US counties. Cohen and his colleagues explored many possible explanations including variations in smoking prevalence, errors in the radon data, coincidental properties of data sets, and confounding by socioeconomic factors, but none of these led to substantial reductions in the effect. In spite of extensive efforts, the negative correlation between radiation exposure and lung cancer rate therefore remained. Consequently "there is no evidence in this analysis that low-level radiation causes cancer, and there is at least some evidence that it may protect against cancer."

On the other hand, the commonly assumed model for radiation protection is the linear- no-threshold (LNT) model, which assumes a direct linear extrapolation from the deleterious health effects of radiation at high doses to the very small doses in our background. The contrasting model of negative associations between low exposures and health consequences giving non-linearity (i.e. a U shaped curve) is referred to as hormesis.

Non-linearity is typically manifested by reduced cancer incidence and increased longevity at low radiation exposures. This is to be expected since organisms evolve over time to be fittest in the environments in which they live, that is exposed to background radiation. This fundamental tenet of evolutionary biology shows that the LNT model is a biological impossibility. In accord with this reality, radiation hormesis has been demonstrated on numerous times in controlled experiments in organisms ranging from microorganisms to mice.

Furthermore, the multiplicity of environmental agents to which we are exposed on Earth interact metabolically, so that the LNT model breaks down to include the exposures of geological outliers including HBRAs as explained in detail in the next section. Therefore the hormetic zone apparently extends to exposures substantially in excess of those at sea-level.

Some French scientists recently carried out a comprehensive assessment of the evidence for and against the LNT model. In an excellent report "Dose-Effect Relationships and Estimation of the Carcinogenic Effects of Low Doses of Ionizing Radiation," published by the French Academies of Medicine and Science, Tubiana *et al.* concluded that the LNT model for human data defaults at doses less than 100 mSv, and even more so at extremely low doses less than 10 mSv. Positive health consequences of low exposures to ionizing radiation are implied compared with the LNT model. Such positive health or fitness deviations from the LNT model are a manifestation of radiation hormesis.

In preparing the French report, the authors note the very real difficulties in obtaining accurate estimates from demographic data. They emphasize that hundreds of thousands of subjects need to be included and monitored for a sufficiently long time, but that "below 20 mSv generally encountered within the context of radioprotection, epidemiology can neither confirm nor refute the existence of an increased incidence of cancer."

The French report is based upon an analysis of a range of populations including the survivors of the Hiroshima and Nagasaki bombs, people involved in the clean up following the Chernobyl accident, nuclear industry workers, and radiologists and radiology technicians. Even airline flight crews are included where the French report concludes:

Airline flight crews receiving exposures of 1.5 to 6 mSv per year have been studied. No increase in the total number of cancers or of cancers in the most radiosensitive organs has been detected in 44,000 members of flight crews or in 2,749 Canadian pilots. An excess of melanomas was observed in these populations, and this can be explained by their more frequent exposure to the sun.

A potentially safe exposure range substantially above those recommended by the International Commission of Radiation Protection (ICRP) emerges from the French report. The lack of deleterious consequences from background exposures in HBRAs, plus the evidence for hormesis, appears consistent with a limit to the safe range in the region of 50-100 mSv per year for the general population.

Therefore, the limit of 1 mSv per year recommended by the ICRP appears absurd in view of these estimates. From this descent into fantasy based upon the demonstrably invalid LNT model, spurious estimates of risks follow. A consequence of these spuriously inflated risk estimates is media hysteria based upon radiophobia for the Chernobyl accident and exposure to depleted uranium.

In the case of Chernobyl, a recent report of the Chernobyl Forum made up of eight UN specialized agencies, has shown that the Chernobyl incident was not catastrophic and not nearly as substantial as had at first been feared. Even so, one prediction of the Forum is that eventually an extra 4,000 may eventually die from cancer. This speculative upper limit is based upon the LNT assumption. In any case, such a predicted number cannot be verified, and equally, it is difficult to falsify. Most importantly, any hormetic model would necessarily imply substantial reductions below the estimated 4,000 deaths, almost certainly closer to zero than to 4,000.

LNT MODEL INVALID BEYOND BACKGROUND EXPOSURES

On evolutionary grounds, organisms tend to become increasingly adapted to their habitats. While fitness is characteristically assessed at the organismic level for a range of traits such as fecundity, survival and longevity, energy provides a more fundamental foundation for fitness. A tendency towards increased energetic efficiency of organisms in their habitats therefore represents an adaptive process which translates into increased fitness, however measured. That is, a basic measure of fitness in a stressed world is energetic efficiency in the face of multiple environmental hazards, including ionizing radiation.

In this context, stress is an energy drain and can be expressed by a loss of fitness. Consequently, stress is an environmental change or probe that targets energy carriers so that energetic efficiency is reduced. Any increase in stress will disturb the energy balance between input and output that organisms build up in their habitats.

More generally, the second law of thermodynamics requires that any process underway in a system universally degrades the energy in that system. Stress therefore reduces energy flow until a critical threshold is reached where cells and hence organisms can no longer survive. Based upon this reductionist approach, the availability of energy and its interaction with stress underlie fitness, and hence can determine limits to adaptation in populations of the past and present.

Variation in the severity of an environmental stress gives a non-linear fitness continuum. Considering temperature, fitness should be maximal at intermediate temperatures between the extremes of heat and cold. Physiologically, this expectation is manifested by low metabolic rates at intermediate temperatures. For example in fasting rats, minimum heat production and maximum survival occur around the intermediate temperatures of 28–29°C. This and other examples show that biological systems are the most stable and energy dissipation is low in the thermal neutral zone, which is a region of high energetic efficiency and hence high fitness.

Since fitness in relation to stress level is non-linear, the LNT model often assumed between the level of an environmental agent and its biological and health consequences is invalid, since all environmental agents have energetic costs. The term hormesis can be used to describe the maxima especially for agents that are exceedingly toxic at high exposures. Radiation hormesis becomes merely a case study of a general ecological and evolutionary expectation for all environmental agents. An array of mild stresses including cold, heat, physical activity, irradiation and caloric restriction can induce longevity extensions. One metabolic consequence of such hormetic exposures to mild stress is the production of "heat shock" or "stress" proteins, hsps. However, this adaptive process can incur a cost, since *Hercus et al.* increased life span of Drosophila melanogaster by repeated exposure to mild heat stress but fecundity fell.

Sørensen *et al.* argue that the expressed hsps level in each species and population is a balance between benefits (resistance to stresses) and costs (negative effects on growth, development rate, fecundity etc.). Fitness maxima of hormetic zones therefore should reflect tradeoffs among various metabolic components all directed towards the maximization of energetic efficiency.

Radiation hormesis is commonly observed in experimental organisms and in man at exposures substantially exceeding background radiation levels. Assume that energetic and metabolic reserves such as hsps are built up to counter the wide array of stresses, especially climatic, to which all organisms are exposed in natural habitats. Such adaptive responses could provide protection from low-level to moderate stresses such as ionizing radiation at non-catastrophic levels but at higher levels than common background.

That is, the hormetic response becomes part of a general stress response involving hsps and other metabolic adaptations across stress levels and environmental agents. Fitness interactions for combinations of stresses should therefore ameliorate the effects of individual stresses in isolation. Therefore the hormetic model can be extended to multiple environmental agents. Cross-protection among various environmental agents could then occur, so that hormesis depends upon the energetic consequences of the totality of interacting environmental stresses of natural habitats. Hormesis therefore is an expression of high energetic and metabolic efficiency and hence high fitness that evolve in response to single and multiple environmental agents where energetic costs are not excessive. This generalization is consistent with the lack of deleterious consequences of life in HBRAs, and to negative associations between radiation exposure and cancer incidence, which only change to positive when exposures become extreme.

OXIDATIVE STRESS, METABOLIC BALANCE AND ADAPTATION

Recent attention given to oxidative stress in natural populations will now be considered in the light of the above energetic and metabolic efficiency arguments. The free radical theory of aging, based on the premise that a single common process is responsible for aging and death of all living beings, is a convenient starting point. Life apparently evolved spontaneously from basic chemicals formed by free radical reactions largely initiated by ionizing radiation from the sun.

Under the free radical theory, the process of aging is determined by the sum of the deleterious free radicals occurring continuously throughout the cells and tissues of organisms. Free radicals, however incited, are proposed to underlie the progressive deterioration of biological systems over time due to their innate ability to produce change. In particular, oxygen is combined with enzymatically degraded food products to produce energy during tissue respiration at the mitochondria. A number of reactive oxygen species, ROS, are generated during this process.

A wide range of cellular molecules including membrane lipids, proteins and DNA, can be attacked and seriously damaged by ROS. These free radicals therefore generate oxidative stress so reducing fitness and accelerating aging. Genetically-determined resistance to a variety of stresses in many species, including yeast, nematodes and Drosophila, is strongly correlated with longevity. Multiple-stress screening including oxidative stress, is an effective procedure for identifying longevity genes.

Similar patterns of gene expression characterize aging and oxidative stress in D. melanogaster, suggesting that the primary or operative feature is oxidative-stress resistance. Individuals with the potential for a long life should therefore carry genes for resistance to ROS, which are an inevitable consequence of life in a world rich in oxygen. In D. melanogaster, the stresses tested include temperature extremes, desiccation, anoxia and ionizing radiation. Recently, three ecological stresses (aridity, high temperature, solar radiation) which differentiate the north and south facing slopes of Evolution Canyon in Israel have been reduced to oxidative stress responses in field collected yeast, Saccharomyces cerevisiae, as anticipated from temperature-stress experiments in D. melanogaster.

Physical stresses in natural populations therefore appear reducible to oxidative stress responses, deriving from the universality of ROS. Oxidative stress responses are therefore critical in the adaptation of natural populations, so that survival becomes a function of the efficiency of handling such stress over time. For example, oxidative stress is lower in healthy centenarians than in other aged subjects. One of the most radioresistant organisms on Earth is the red-pigmented bacterium, Deinococcus radiodurans, which is resistant to other physical stresses including ultraviolet radiation, hydrogen peroxide, heat, desiccation and a variety of toxins. Genes protective against radiation stress are therefore protective against other stresses. For example, antioxidants can protect against both X-irradiation and oxygen poisoning in mice.

Damage from increased oxidative stress can be related to an imbalance between free-radical production and antioxidant protection. While this imbalance strengthens as radiation exposure increases, it should be minimal where fitness is highest in hormetic zones in accord with evolutionary expectations. A range of physical stresses should increase this imbalance, but metabolic interactions among stresses should adaptively ameliorate the consequences of individual stresses so providing metabolic protection. A significant consequence is radiation hormesis for exposures substantially in excess of common background exposures.

Radiation hormesis therefore becomes a manifestation of minimum metabolic imbalance. Analogously considering nutrition, excessive diets elevate the production of ROS implying increased metabolic imbalance, so giving a non-linear relationship and hence nutritional hormesis. In any case in the hormetic zone of caloric restriction, metabolic rate is less and stress resistance is higher than under excessive diets, implying a region of high metabolic balance.

Radiation hormesis can be viewed as the summation of interdependent and complex adaptations involving ROS, antioxidants, heat shock proteins, DNA repair processes, and so on. More generally, complex metabolic processes at the molecular, genetic, chromosomal, cellular, physiological and immunological levels are interdependent, and have been considered in the hormetic context by increasing numbers of authors. Hormetic deviations from the LNT model follow, manifested by high fitness at the whole organism level typically measured by longevity and survival, and in demographic data by reduced cancer incidence. The biosystem therefore responds to ionizing radiation so efficiently in the hormetic zone that the risk of mortality from cancer should fall and life span should increase due to protective metabolic processes, apparently reducible to the evolutionary consequences of oxidative stress.

THE EVOLUTIONARY EXPECTATION INVALIDATES THE LNT MODEL

The invalid LNT premise has been documented in over 5,000 examples of physical/chemical agents. Scattered observations back to the early part of last century are even suggestive of hormesis in bacteria for uranium salts, and in the last decade a low-dose hormetic response to depleted uranium in grass has been reported. More information would assist, but uranium is a constituent of most rocks and the sea, giving an evolutionary expectation of uranium hormesis.

It is not surprising that Gerber *et al.* concluded: "We suggest that hormesis may be almost universal for substances normally present throughout geological time." Habitats provide the ecological theatre for assessing the LNT model. Adaptation towards high energetic and metabolic efficiency, that is fitness, to counter the metabolic consequences of the stresses from single and multiple environmental agents should occur over time. This evolutionary process is expected to lead to non-linear continua for fitness for all agents including ionizing radiation.

In the specific example of radiation, hormesis extends to exposures substantially in excess of background due to interactions between the metabolic effects of an array of abiotic stresses to which organisms are exposed in their habitats. Non-linearity should therefore be the primary hypothesis against which any data set is tested. Indeed, Calabrese has said "I actually think that there should be a paradigm shift and that the hormetic model should be the default model." This quote comes from Hadley, one of the few commentators who has considered the need to emphasize evolutionary inputs in attempting to understand hormetic zones. She emphasizes the underlying evolutionary process of selection for metabolic efficiency to adapt to a stressed world, which immediately predicts the universality of hormesis for all environmental agents. This is a present-day version of the often repeated Dobzhanky dictum that "Nothing makes sense in biology except in the sense of evolution." In conclusion, non-linearity is the fundamental model for assessing the risks from lowdose radiation.

PROTECTION OF THE ENVIRONMENT FROM IONISING RADIATION

The need for a system to protect the environment from ionising radiation is now recognised and there has been a considerable international and national effort on this issue over the last decade. The focus has been on collating relevant information and developing approaches to enable regulatory assessments with the aim of demonstrating whether the environment is adequately protected from the exposure to ionising radiation released under authorisation.

The key is to be able to demonstrate that the environment is protected. Whilst many people believe that the ICRP statement ('the Commission believes that the standards of environmental control needed to protect man to the degree currently thought desirable will ensure that other species are not put at risk') continues to be valid, most accept that it is difficult to explicitly demonstrate that the environment is protected using the original ICRP statement. Given developments in conservation objectives and legislation, there is now a need to be able to demonstrate environmental protection (e.g. as the EC Habitats Directive has been interpreted and implemented in England and Wales).

Concerns have been expressed both from the nuclear industry and the scientific community that the developing approaches may lead to costly regulation. However there has been a considerable amount of work funded by non-industrial sources to develop tools and approaches that can be used for the purpose of environmental assessment. Some of these tools are freely available can be used by industry to provide assessments to their regulatory bodies. Furthermore, it is recognised that the potential requirement for, and consequences of, regulation should be weighed against the possible role that nuclear power generation may have in (e.g.) combating climate change. Being able to explicitly establish whether radioactive substances are/are not having an environmental impact will contribute to the debate of the use of nuclear power in the future.

It is noted however that validation and comparison of the radioecological and dosimetry components of various approaches is required and this has also begun through the Biota working group of the IAEA's EMRAS programme. However, it is also important that any approaches used are practicable, credible to stakeholders and fit for purpose in any regulatory context. The EU Euratom funded Protect project is evaluating different approaches to protection of the environment from ionising radiation and will compare these with the approaches used for non-radioactive (chemicals) contaminants. The project aims to provide a scientific justification on which to propose numerical targets or standards for protection of the environment from ionising radiation which could be used in a regulatory context to aid in demonstrating whether the environment is adequately protected. To achieve this aim, the Protect project has consulted widely within Europe, and the broader international community, with industry and regulators. This paper will report on the critical review that has been conducted within Work Package 1 of the Protect project.

METHODS

The first stage of the Protect project has been dedicated to reviewing national and international regulatory methodologies and criteria currently applied to environmental protection from radioactive substances. This included assessing the regulatory instruments, procedures and underlying principles, and criteria currently applied in different countries. Environmental regulators, nuclear and non-nuclear industries, international organisations and NGO's were asked to identify the key regulatory instruments for assessment and give their views on how environmental regulation is applied.

Out of approximately 130 organisations contacted, questionnaires responses were received from 50. Regulators and industry each constituted 36% of the respondents, NGOs and international organisations 10%, and advisory bodies 18%. Although the questionnaires were primarily targeted at environmental regulators and representatives from industry within EU member states it was recognised that worldwide perspective would also be valuable and responses were also sought and obtained from, for example, Canada and Australia.

The review assessed similarities and differences in approaches for chemicals and radioactive substances. It evaluated the extent to which the existing approaches fulfil the objectives of environmental protection by looking at what endpoints are being applied, what is acceptable in terms of permitted risks, what levels of compliance are required for chemicals and radioactive substances (and are there any differences) and are there common themes in the application of approaches for chemicals and radioactive substances.

Historically, there has been a different philosophy for regulation of radioactive substances and chemicals. Regulation of radioactive substances has been focused on the protection of man with the environment as a secondary consideration whereas for the regulation of chemicals the environment has traditionally been a focus.

Therefore it is important for developing radiological environmental protection to learn from the evolution of chemical practices, bearing in mind any broader environmental protection objectives, and to negate any confusion that may currently exist between chemical and radiological environmental protection.

Radioactive substance regulation with respect to environmental protection is not currently defined by international or European legislation. In contrast, regulation of chemicals is usually though the application of internationally accepted standards. A similar approach for radioactive substances could be seen as beneficial as such regulations across Europe would facilitate the centralisation of knowledge and effective planning and this may be the end result of current and future developments in the field of environmental protection from ionising radiation. Harmonisation between approaches for chemicals and radioactive substances was considered desirable; particularly as a similar level of protection for both using common assessment endpoints where possible would have a number of advantages. For example, a series of common measurement endpoints, which could be normalised to risk, would allow comparison of the relevant impact from different contaminants.

It was recognised that there are practical reasons for differences in terms of how criteria to demonstrate protection for chemicals or radioactive substances are set. For example, criteria for radioactive substances are consolidated, or summed, to include all radionuclides and exposure pathways, whereas in most cases, criteria for non-radiological contaminants are based on single contaminants and exposure pathways. It is clear that the same basic risk assessment paradigm applies to both radiological and chemical assessment. Essentially, it involves comparing an estimate of exposure to some tolerable level or dose. If the actual or expected exposure is greater than the tolerable level, then this might trigger some action, or at least prompt more accurate exposure estimates. A key element within these risk assessment schemes is the need for thresholds that define acceptable levels of stressors which is considered further by Andersson *et al.*, (these proceedings).

The basic generic framework for both radioactive substances and chemicals is: Problem Formulation, Exposure Assessment (e.g., emission estimation tools, dispersion models and food-chain models), Effects Assessment (e.g., estimation of tolerable concentrations in the environment), Risk Characterisation and Risk Management. These steps are similar to those covered by the more developed radiological assessment tools (e.g., the ERICA Integrated Approach. However, this was intentional as practices in chemical assessment were considered in the development of these radiological assessment tools.

The key recommendations that have come from the work are that:

- Environmental protection goals are often stated in a country's legislation but these are often broad ranging and aspirational in nature using, for example unspecific terms such as: 'to protect the environment', 'to protect ecosystems' or to maintain 'favourable conservation status', 'biological diversity', 'structure and function of habitat/ecosystem', 'protected species' or 'rare species'. Following detailed discussions it was determined that protection should focus on the population level and that protection goals should be translated into measurable targets with advice provided on what the tolerable risks associated with these endpoints should be. Endpoints that relate stressor levels to measurement endpoints such as morbidity and reproduction should be targeted because ecological theory shows that these traits determine population sustainability. A caveat is that individuals (e.g. those that are rare or endangered species) may need to be considered specifically.

- It is clear that there is a strong advocacy for linking radiological protection to the processes used for chemicals assessment. Although there are some technical differences, the underlying protection goals are similar and broadly the same risk assessment approaches may be used. For example, the use of Species Sensitivity Distribution and Assessment Factor approaches to determine benchmark dose rates based on agreed tolerable risks should be encouraged and the use of purely expert judgement should be avoided where possible.
- The use of the numeric (dose rate) values currently being applied, or suggested, should be assessed and the need for screening values and standards considered. Where possible, harmonisation of future international guidelines and recommendations should be attempted (for example, IAEA Basic Safety Standards, EU-Directives, and ICRP Recommendations).
- Protect should produce a clearly understandable document outlining the derivation of any numeric benchmark values, in particular explanation of where there are limitations in the application because of poor data quality is needed and the level of conservatism in the values should be clearly identified. This document should be developed in consultation with stakeholders.
- Furthermore, once numeric values or some other methods of environmental protection have been agreed, methods for demonstrating compliance should be evaluated (bearing in mind the use of the numeric value for example, if a regulatory limit then clearly strong compliance will be needed).
- Optimisation of discharges should remain central to environmental/human radiological protection.
- The benefits of regulation for the nuclear and non-nuclear sectors should be identified as they are likely to demonstrate that users of radioactive substances are behaving in an appropriate and environmentally responsible manner. Emphasising and highlighting this could be beneficial in terms of large scale environmental issues such as climate change with respect to nuclear power generation.

Further details are available in the final deliverable report. The information in this report is contributing to other PROTECT work packages to determine the appropriateness of numeric benchmarks suggest for use to demonstrate environmental protection from the effects of ionising radiation and to address the issues of the cost, or burden, of any environmentally derived regulation (Beresford *et al.* and Andersson *et al.* these proceedings). The overall outputs from the PROTECT project will contribute to the debate on the environmental consequences of nuclear power generation and radioactive waste disposal.

In summary, assessment tools have been developed that allow environmental assessments to be conducted and which can be used to demonstrate whether non-human species are protected, or not, from the impact of exposure to ionising radiation released under regulation. There is still work to do to evaluate whether these approaches are fit for purpose and this is underway. Frameworks are often tiered, increasing in the level of complexity, input data requirements and need for expert involvement. These are conceptually based on, assessment approaches available for chemical risk assessment. What criteria, and how these should be defined, for use in tiered assessments is currently under debate but are likely to make use of approaches developed for setting assessment criteria for chemicals. However it will be necessary to ensure that any assessment criteria that are developed are fit for use in a regulatory context.

ADAPTIVE RADIATION OF ISLAND PLANTS

The presence of diverse and species-rich plant lineages on oceanic islands is most often associated with adaptive radiation. Here we discuss the possible adaptive significance of some of the most prominent traits in island plants, including woodiness, monocarpy and sexual dimorphisms. Indirect evidence that such traits have been acquired through convergent evolution on islands comes from molecular phylogenies; however, direct evidence of their selective value rarely is obtained.

The importance of hybridization in the evolution of island plants is also considered as part of a more general discussion of the mechanisms governing radiations on islands. Most examples are from the Hawaiian and Canarian floras, and in particular from studies on the morphological, ecological and molecular diversification of the genus Aeonium, the largest plant radiation of the Canarian Islands. Why are some plant lineages more speciesrich than others? The question arises repeatedly when studying the floras of isolated oceanic islands. Within these restricted areas, numerous plant genera contain a considerable number of endemic species. Such island radiations are of particular interest because, unlike many continental plant groups, they have radiated within a very restricted geographic area and their relatively recent, rapid radiation can be demonstrated through the geological history of the islands. A radiation is defined as a process leading to character diversification and/or ecological diversification and eventually may lead to lineage diversification.

In this paper the focus will be on radiations leading to lineage diversification. We restrict the concept of radiation to cases in which diversification occurs at an accelerated rate, as was also done by Sanderson and Bateman, though some authors avoid this more restricted view of radiation. An adaptive radiation is defined as the diversification of a lineage into species that exploit a variety of different resource types and that differ in the morphological or physiological traits used to exploit those resources.

It is generally assumed that most cases of island radiation, including the examples presented here, are adaptive in this sense. Whether plant radiations are a more frequent phenomenon on islands and island-like habitats than on continents still needs to be investigated. It is possible, though, to list a number of deterministic and stochastic forces that generally promote speciation and that may be particularly prominent on oceanic islands. Determining the relative importance of these forces is a difficult task, especially since they may vary between lineages and islands. The rich diversity of habitats on many oceanic islands, created primarily by geological and climatic conditions, offers a wide array of ecological opportunities to colonizers.

In addition, due to the remoteness of some islands, the number of organisms capable of colonizing from the nearest continent is limited, and competition and predation from well-established and specialized plants and animals may be low. Under these circumstances, selection may favour characters which offer the opportunity to exploit new resources and which eliminate competition with close relatives. Sometimes the acquisition of a particular trait or traits predisposes a taxon to rapid speciation, thus giving access to a new range of resources.

For example, the evolution of the spur in continental Aquilegia (Ranunculaceae) appears to be such a key innovation which led to rapid radiation within the genus. Other circumstances may also favour radiation: in his flush-crash-founder model Carson suggests that episodic relaxation from selection may cause disruption of old adaptive gene complexes and allow selection to operate on new genetic systems.

Random evolutionary processes are also invoked to explain island radiations. With habitat turnover due to volcanic activity, and the opportunity for repeated cycles of inter-island dispersal and isolation, population bottlenecks, founder events and geographic separation of conspecifics may have occurred repeatedly through the evolution of island floras. These factors will promote evolution by drift; this could occur, for example, through the random fixation of different advantageous genes in separate populations experiencing similar selection pressures.

Not all lineages on an island or archipelago are equally successful, even when they have the same ecological opportunities and are exposed to the same chance events. This may be a simple question of colonization order, or may reflect chance events which lead to extinction. Variation in the success of lineages may also be due to differences in the evolutionary potential (or genetic variability) and the phylogenetic constraints (or genetic composition) between lineages. Table lists the genera in the Hawaiian and Canarian archipelagos comprising more than ten endemic species. It is apparent that genera within the Asteraceae have been particularly successful in diverging within both archipelagos, and that the family Campanulaceae has several species-rich genera in the Hawaiian archipelago.

This pattern of several independent radiations within the same families suggests a phylogenetic component to island radiations. However, it may also reflect factors such as the presence within these families of a large and diverse continental species pool or advanced dispersal abilities that ensured more frequent colonizations of the islands. Soon after establishment on an island the dispersal capacity may be lost rapidly. Species richness and the presence of morphological and ecological diversity in a lineage are often taken as evidence for adaptive radiation in island plants, in particular where this is combined with low genetic differentiation.

In such rapid, adaptive radiations, neutral characters are expected to exhibit little diversity while features related to fitness will vary considerably. Few studies, though, actually seek to identify the significant traits in these suggested adaptive radiations. In this paper, we assess this problem by discussing the possible adaptive significance of some of the most prominent traits in island plants. The importance of hybridization in the evolution of island plants is also considered as part of a more general discussion of the mechanisms governing radiations on islands. Most examples will be taken from the Hawaiian and Canarian floras and in particular from our studies on the morphological, ecological and molecular diversification of the Canarian genus Aeonium. Our choice of examples reflects both the geographic bias in the literature of island plant radiations and the fact that the flora of these two archipelagos has been very well recorded. Both archipelagos are of volcanic origin but differ in age and degree of remoteness. The Hawaiian Islands are situated some 4000 km off the Californian coast and comprise eight major islands varying in age from 0.5 to 7.5 Mya. The seven major islands of the Canarian archipelago lie some 100 km from the African coast and range in age from 0.8 to 21 Mya. The succulent, rosette-forming species of the genus Aeonium (Crassulaceae) comprise the largest radiation in the Canary Islands.

Thirty-six of the 42 species in the genus are endemic to the archipelago while the remainder are distributed on other Macaronesian islands (two species on Madeira and one species on the Cape Verde Islands), in Morocco (one species) and in East Africa (two species). The genus probably arose in the Canarian archipelago and later dispersed to Madeira, Cape Verde and Africa. All of the seven major islands of the Canarian archipelago are inhabited by members of the genus Aeonium. Most species are single-island endemics and confined to particular habitats of an island. Various authors regard the genus as a prime example of adaptive radiation in plants due to the morphological diversity in the genus, the restricted habitat use of most species, and the wide array of habitats exploited by different species.

ARE PLANT RADIATIONS IN REPRODUCTIVE OR VEGETATIVE TRAITS?

To obtain an accurate reconstruction of character evolution within the groups requires well-resolved phylogenies, but such information is not available for these groups. However, a crude estimate of the relative amount 32 T. H. Jorgensen and J. M. Olesen of variation in reproductive and vegetative characters in plant groups can be obtained by counting the number of dichotomies in the taxonomic keys to these groups. This approximate method is likely to be biased towards the most obvious and apparent morphological traits of the plants. In Table we present the counts of vegetative and reproductive dichotomies for the most species-rich genera (>10 endemic species per genus) in the Hawaiian and the Canarian Islands using the flora of Wagner *et al.* and Bramwell and Bramwell.

It appears from this analysis that many genera have approximately equal numbers of vegetative and reproductive dichotomies. In the Canary Islands, though, there is a bias towards dichotomies in vegetative characters. Genera with vegetative dichotomies are assumed to have flowers and pollinator faunas that scarcely vary among their species. The pollinator faunas may also be random samples of the local pollinator community. In the Hawaiian Islands it is mainly the genera pollinated by drepanid birds that exhibit reproductive dichotomies. This is because, following dispersal and isolation, species pollinated by only one or a few pollinator groups are more prone to differentiate in their floral structures compared to those visited by a random array of pollinators. In the case of Aeonium, only 20% of the dichotomies in the key are floral, suggesting that this genus has a rather randomly assembled pollinator fauna. A study of A. holochrysum on Tenerife supports this argument. Preliminary data from two study sites show that a large part (48% and 75%) of the species observed to visit the flowering plants in the habitats (Coleoptera, Hymenoptera, Diptera and Lepidoptera) visited the flowers of A. holochrysum.

The flowers of Aeonium conform to the general Crassulacean type by being radially symmetric, hypogynous, having a corolla differentiated into petals and sepals, and having two whorls of stamens and one whorl of nonfused fruits in the form of follicles. Nectariferous glands are formed at the base of each carpel. Flowers are yellow, white or reddish. Inflorescences are panicles with a variable number of branching orders. Taken as a whole, the genus appears very uniform in terms of floral shape, and this is reflected in the predominance of vegetative dichotomies in the taxonomic key.

There is, however, extensive variation in size of flowers and inflorescences, and a detailed investigation of flower morphology in 18 species revealed subtle differences between species in shape of floral organs and inflorescences. Correlating distances in species means of these traits to their patristic distance based on the molecular phylogeny of the genus in a partial Mantel test revealed that variation in various floral characters – size of flowers (here: petal length), pubescence of floral organs, and number of petals—has a substantial phylogenetic component.

This provides evidence of some differentiation in reproductive traits during the early radiation of the genus despite the constraints to evolution imposed by the Crassulacean flower architecture. Pollen and ovule production, two important components of the reproductive output of a plant, vary to some extent among species of Aeonium. Pollen-ovule ratios in the range of 92-313 might lead us to the conclusion that Aeonium species are facultatively autogamous.

Both the number of ovules per ovary and the number of pollen grains per stamen are variable among species. Other components of the reproductive output in Aeonium species (number of ovaries and stamens per flower, number of flowers per inflorescence) vary similarly. For example, the monocarpic species (A. urbicum and A. hierrense) have a considerably higher flower production per inflorescence than closely related polycarpic species. Of the four components listed in Table only the numbers of ovaries and stamens per flower appear to have a phylogenetic component.

We suggest, therefore, that a part of the variation in the four components is due to present ecological conditions selecting for a fine-tuning of reproduction to specific habitat requirements by altering various components of the total reproductive output. It awaits further investigation of reproductive traits in more Aeonium species and, in particular, of the ecology of all members of the genus to understand fully the importance of these traits in the radiation of the genus. Other island genera without prominent differentiation in floral traits also need to be investigated. Characters related to growth form contribute significantly to the numerous dichotomies involving vegetative traits in the taxonomic key of Aeonium.

The 42 species of Aeonium vary in appearances from creeping and sparsely branched, or upright, unbranched woody forms to subshrubs or tall shrubs. Leaf morphology is also an important trait in the taxonomic classification of species. The rosette leaves vary in shape, pubescence and presence of glands and ciliate margins. Molecular phylogenetic investigations imply that the evolution of the different growth forms in Aeonium is related to early events in the radiation of the genus. A detailed analysis of the growth forms of 22 species of Aeonium additionally reveals that, not only do closely related species end up with the same growth form, but they also reach their general appearance by the same form of modular growth.

Simple linear and logistic regression models can be developed to predict module length, branching mode and flowering probability in species of Aeonium based on their modular construction. When combined, the parameters of these models are species-specific and describe the growth rules of each species. Species are separated significantly at the level of phylogenetic section in a discriminant analysis based on the parameters of these growth rule models. Since species with the same growth rules are confined to the same phylogenetic clade, we propose that early evolution of these growth rules has been important for further radiation in the genus.

ARE RADIATIONS ADAPTIVE?

What is the relative importance of adaptive and non-adaptive evolutionary processes in the radiation of Aeonium? This issue can be assessed indirectly by comparing the phylogeny of a plant group with the geographic and ecological distribution of the various species. Species of Aeonium in the same clade and with the same growth rules often occur on different islands but in similar kinds of habitats. It is suggested that dispersal across similar habitats of different islands, followed by divergence due to isolation, is a prominent route in speciation in Aeonium. The same phylogenetic pattern of dispersal and diversification has been well documented in another Canarian genus, Argyranthemum, and may also be valid for other large genera in the archipelago, e.g. Sonchus and Crambe.

In other island groups, e.g. the Hawaiian silverswords alliance and Cyanea, the radiation of a lineage into new habitats within the same island appears to be more significant than divergence due to isolation after colonization of similar habitats on different islands. A strong indication of the adaptive value of a trait is when phylogenetically separate, but ecologically similar, species converge or show parallel patterns of variation along similar ecological gradients.

In Canarian lizards, for example, the correlation of changes in colour pattern with environmental variation is parallel in different species on neighbouring islands, and is taken as evidence of natural selection. In Aeonium, similar analyses have been done for four unbranched species within the section Leuconium. The four species are widely distributed within both mesic and dry habitats of the four western islands. By the use of partial Mantel tests (and correcting for phylogenetic relatedness), it was revealed that traits related to plant height, leaf shape and flower morphology vary predictably with distance from continental Africa.

The result implies that variation in these morphological traits is due to different degrees of "oceanity" and may be a result of selection. Different degrees of oceanity may reflect decreases in temperature amplitude and solar radiation with distance from the continent. To investigate in more detail the adaptive significance of plant traits would require functional analyses or replacement experiments, such as those performed in animal radiations like Caribbean lizards or sticklebacks. A number of distinctive traits occur repeatedly in island plants.

Some of the most frequently mentioned traits in the literature of adaptive radiation on oceanic islands are the woody habit, monocarpy, developmental heterophylly and sexual dimorphism. Before concluding that these traits are particularly favourable in an island environment and facilitate adaptive radiation, it it necessary to determine their place of origin. It is reasonable to suppose that a trait, which is acquired more frequently in island lineages (and is found in different archipelagos) than in continental groups, is of selective value in the island environment and has island origin. Alternatively, a trait may be common in island floras when it has facilitated the radiation of single lineages. In the latter case, though, we will be left with the difficulty of determining whether such a trait is a key innovation or whether it is a nonadaptive trait in a lineage that has radiated due to the general availability of open, unsaturated habitats. Support for the hypothesis that a particular trait facilitates the radiation of island lineages could be obtained by comparing sister groups exhibiting the presence and absence of the trait.

Ideally, this test should be based on a detailed knowledge of phylogenetic relationships and the timing of speciation and extinction events. Such a complete knowledge appears to be lacking for island plant groups and their continental sister lineages. The following discussion on key features thought to be involved in the adaptive radiation of island lineages are therefore not yet confirmed statistically.

Woodiness

Woodiness is a prominent trait in island endemics. In the most species-rich genera of the Canarian and Hawaiian Islands the large majority of species are woody. But is this trait adaptive and has it facilitated diversification within islands? – Hypotheses as to why woodiness has evolved frequently in the insular condition are numerous but yet untested. It could be a non-adaptive response to a stable environment or an adaptive response to the release from seasonality. If the latter is true, it could be the result of selection for increased stature improving competitive ability with other plants or the result of selection to increase longevity in a pollinator-scarce environment.

Yet another suggestion is that release from large herbivores permits delayed maturity and reproduction . Insular woodiness is an acquired trait in the Hawaiian silverswords, i.e. the lineage comprising Argyroxiphium, Wilkesia and Dubautia, and in Canarian Echium and Sonchus. It is also likely that the North African ancestors of the monophyletic group comprising Aeonium, Aichryson and Monanthes were herbaceous, though the majority of species in these lineages are woody. In Macaronesian Sempervivoideae there is an increase in woodiness from the most ancestral Aichryson, which includes annual species, to the most derived Aeonium, which are all woody.

It is possible that woodiness was a key feature in the radiation of these lineages. The pattern is not consistent, though, as other island genera appear to have a woody continental ancestor; many others remain to be investigated.

MONOCARPY

The unbranched, monocarpic form has evolved at least four times in the genus Aeonium, and several times in the Hawaiian Cyanea. It is also common in other plant groups of oceanic islands or island-like habitats (e.g. Canarian Echium and East African Lobelia). Factors favouring monocarpy may be of a demographic nature where ecological conditions lower the "value" of an adult compared to that of the juvenile.

The disturbed and barren areas such as lava flows and abandoned fields, which are among the habitats of unbranched Aeonium urbicum, A. appendiculatum, A. hierrense and A. pseudourbicum, may be hypothesised to favour short life spans and therefore monocarpy. Alternatively, monocarpy may evolve when greater benefits of reproduction come at higher levels of reproductive effort. This could apply to the pollinator-scarce environments of Wallace, if pollinators prefer large inflorescences and fruit set increases relatively with inflorescence size. As with the demographic models this suggestion awaits further testing in island plants.

HERBIVORE DEFENCES

Traits related to herbivore defence appear repeatedly in different island lineages. The fossil record reveals the historic presence of large browsers like the "moa-nalo" ducks and giant geese on the Hawaiian Islands, the flightless moas in New Zealand, and the giant tortoises in the Canarian, Mascarene and Galápagos Islands. Tortoises are still present in the latter archipelago.

As the only large browsers present on the islands (as deduced from the fossil record), it is apparent that these animals have exerted a strong and uniform selection pressure in contrast to that of a diverse continental herbivore fauna. The advantage of evolving anti-predator traits on these islands is obvious and may have faciliated radiations of island genera through repeated evolution in separate lineages or as a key innovation. The high incidence of developmental heterophylly (i.e. juvenile leaves different from adult leaves) in some island floras may be the result of selection for traits that deter herbivores through visual mimicry or visual divergence.

Heterophylly is a widespread trait in the Mascarene Islands and in New Zealand. In the Hawaiian Cyanea, 17 of the 52 species are heterophyllous and the trait has evolved independently at least four times in the genus . The trait coincides almost exclusively with the occurrence of prickles on stems and/or leaves which strongly suggests that it is an anti-herbivore defence.

The original suggestion was that prickles deter herbivorous endemic snails, whereas others have argued that large avian browsers were the primary herbivores. The genus Aeonium has no apparent morphological traits related to herbivore defence, though it remains to be investigated whether any biochemical traits are of significance in deterring herbivores.

SEXUAL DIMORPHISMS

Nearly 21% of the native species in the Hawaiian flora are sexually dimorphic, the highest proportion of any flora studied. New Zealand also has a high incidence of sexual dimorphisms, while the Canaries are similar to the Galápagos in having a low number of such dimorphic species.

All species of Aeonium appear to have sexually uniform flowers. The selective forces suggested to favour dimorphic traits like dioecy, subdioecy, polygamodioecy and gynodioecy involve the promotion of outcrossing. Island plants may suffer from low levels of genetic diversity following colonization and establishment. The evolution of outcrossing mechanisms may therefore favour further radiation and diversification in island groups.

An alternative suggestion is that ecological factors favour dimorphism as a result of altered patterns of resource allocation and sexual selection. The association of sexual dimorphisms with fleshy fruits, woodiness and small, inconspicuous flowers in Hawaiian Islands and elsewhere tends to support this theory. In the Hawaiian Islands, at least 12 lineages have evolved dimorphisms upon establishment in the archipelago.

These have given rise to one third of the current dimorphic species. Within the lineage comprising the 26 species of Schiedea and Alsinidendron, for example, dimorphisms have at least two origins, and further radiation has resulted in 10 gynodioecious, subdioecious and dioecious species.

HYBRIDIZATION

Hybridization and introgression appear to be widespread processes in flowering plants and are thought to be important in plant evolution. Hybridization can lead to either homoploid or polyploid individuals. The evolutionary significance of this process can be either the establishment of new stabilised taxa with morphologically intermediate or distinct characters, or the increase of the genetic diversity within species as a result of new allelic combinations and the possible breakdown of reproductive barriers following backcrosses to either parent. The successful establishment of new hybrid species depends on the presence of unoccupied niches and may therefore be more likely in open, disturbed habitats. This has been suggested as one of the main reasons that hybridization is important in evolution on islands. In the Canary Islands, as well as in other oceanic archipelagos, volcanic activity has repeatedly caused plant populations to become fragmented and new habitats to arise and in this way may have provided opportunities for new species of hybrid origin to establish.

These opportunities may have been further enhanced by the general presence of diverse and often unsaturated habitats. Additionally, the small amount of genetic differentiation between recently radiated species within a lineage suggests a lack of genetic barriers to hybridization in many island groups. A survey of three continental floras and one island flora performed by Ellstrand *et al.* supplemented with a similar survey of the Canarian flora, reveals no indication that hybridization is more frequent on oceanic islands than on continents.

Species of hybrid origin comprise 5-18% of the native and naturalized taxa in the continental vascular floras (excluding intraspecific taxa), while 9% of the Hawaiian taxa and 4% of the Canarian taxa appear to be derived from hybrids. Another important conclusion from the survey by Ellstrand *et al.* is that natural hybrids occur non-randomly over families and genera. This indicates that, in addition to habitat availability, phylogeny also influences the importance of hybridization in plant evolution.

In the large island genera from the Canarian and Hawaiian Islands it is therefore not surprising to find great variation in the number of natural hybrids recorded. It is apparent that species in some genera readily form hybrids, and in these groups hybridization may have had an impact on the evolution and divergence of species. The majority of contributions to the pool of natural hybrids recorded on the Canary Islands is from Crassulaceae, and from the genus Aeonium in particular.

Aeonium may be extreme in its absence of internal reproductive barriers to crossing with congenerics. Natural hybrids are recorded for almost all species occurring in sympatry or parapatry with the exception of those where flowering periods do not overlap. Whereas the occurrence of natural hybrids reflects the potential for the species of a group to hybridize, it does not necessarily reveal the actual importance of hybridization and introgression in the evolution of the group. The occasional hybrids may not become established due to low survival and reproduction. Traditional methods to investigate putative ancient hybrid species have been through morphological comparisons with suggested parent species and their artificial cross products. This is a time consuming method that relies on the existence of extant parental species.

The development of numerous molecular techniques within recent years has created new opportunities for the identification of ancient hybridization. The most informative way is to reconstruct the molecular phylogeny of the group in question by using more than one molecular marker and preferably multiple populations. Because of incongruencies between the phylogenies based on nuclear DNA, chloroplast DNA or morphology, it has been suggested that two species of Aeonium are of hybrid origin.

Morphological characters of A. percarneum and A. lancerottense and chloroplast restriction site data of A. lancerottense suggest their close relationship to species within the section Leuconium. Opposing results come from sequences of nuclear ribosomal DNA (ITS regions) which group A. percarneum and A. lancerottense within the section Aeonium.

If we assume that the chloroplast genome is maternally inherited, these asymmetries are best explained by events of introgression solely via the transfer of pollen from a member of the section Aeonium to A. percarneum and A. lancerottense in the section Leuconium. Evolution by polyploidy appears to be of minor importance in Aeonium as only four species are known only as tetraploids. The strength of the hypothesis of hybridization depends on a number of factors.

Intragenomic processes may render ancient hybridization difficult to detect on the molecular level, though the use of nuclear markers introduces the possiblity of detecting multiple variants in a plant. It is also necessary to keep in mind, when comparing the frequencies of hybridization in different genera, that differences in life histories (e.g. annuals vs perenials) may cause differences in the propensity for hybridization. Investigations that attempt to evaluate the importance of ancient hybridization in the radiation of large island genera are scattered and, as in Aeonium, often incomplete as they do not involve all species of the genus or do not include multiple populations of individual species.

Table lists the molecular studies performed on large genera of the Canarian and Hawaiian Islands with reports on the evidence of ancient hybridization and comments on the relevant methods. All these genera are dominated by perennials. It must be noted that many of these studies were not designed to detect ancient hybridizations. The general conclusion from Table is that where sampling and/or the use of molecular markers has been extensive, evidence of ancient hybridization has usually been found, thus confirming the role of hybridization in the evolution of island genera.

An exception is the genus Echium for which no evidence of ancient hybridization has been found. In Aeonium as well as in other island genera, species often occur only as a single or as a few small populations. It is possible that events of introgression have had a large positive impact on the genetic diversity of these populations through their colonizing history.

However, in small populations, the possible negative effects must also be considered, especially when hybridization is with a more widespread species. In such cases the genotypes of the rare species can be completely absorbed by the greater gene pool of the common species and the integrity of the rare species may be destroyed. With our present knowledge, it is impossible to predict the importance of this factor in the colonization history of Aeonium and other island genera.

Our ability to interpret diversification on islands (and elsewhere) is strongly dependent on an adequate knowledge of the phylogeny of the taxon in question. Molecular data (preferably sequences from non-coding regions) are useful as they are assumed to be independent of morphological variation and avoid the problem of homoplasy. Several revisions of the theories and conventions concerning evolution on islands are based on molecular data.

For example, the repeated confirmation of insular woodiness as a derived character state in Canarian genera leads to the rejection of the former widespread view that these species are relicts and ancestors of many modern Canarian and continental species. Similarly, detailed knowledge of phylogeny will enable us to actually test whether plant lineages on islands are more species-rich than on the continents and whether this is facilitated by the evolution of certain key features. However, there are practical problems.

When traditional molecular markers are applied to island radiations we are most often left with little genetic differentiation compared to that within continental species groups. Future application of marker systems that solve relationships between closely related species is therefore needed.

Radiation and speciation in Aeonium have probably been facilitated by both random and non-random evolutionary processes, as well as by events of hybridization and introgression. From our current knowledge, three factors of an ecological or genetic nature have probably been particularly important in making this genus the largest radiation in the Canarian archipelago. Firstly, our molecular data suggest that dispersal of fruits and seeds may have assured the repeated colonization of different islands, leading to the evolution of new lineages after isolation. Secondly, a lack of barriers to hybridization means that natural hybrids are common, and so introgression may have contributed to formation of new taxa. Thirdly, the most prominent variable trait in the genus, growth form, is determined by simple rules of growth which are probably determined by simple genetic differences. The diversification into new forms may therefore have been easily facilitated by small changes in this genetic background.

Chapter 8

Safe Water Technology for Arsenic Removal

Arsenic contamination of drinking water has been reported from many parts of world. In some arsenic affected areas, substitution of drinking water source by a safe and easily available one may not be possible during part or all of the year, or may be very expensive. Arsenic removal may be a more appropriate water supply option in these situations. This paper describes some safe water technologies for arsenic removal. Coagulation is the most common arsenic removal technology.

As many Bangladesh waters contain arsenite, oxidation with chlorine or permanganate is required first. Coagulation with ferric chloride works best at pH below 8. Alum has a narrower effective range, from pH 6-8. Ion exchange resins are commercially produced synthetic materials that can remove some compounds from water. These resins only remove arsenate. Activated alumina, like ion exchange resins, is commercially available in coarse grains. Activated alumina beds usually have much longer run times than ion exchange resins, typically several tens of thousands of beds can be treated before arsenic breakthrough. Activated alumina works best in slightly acidic waters (pH 5.5 to 6).

Membrane methods for arsenic removal include reverse osmosis and nanofiltration. Currently available membranes are more expensive than other arsenic removal options, and are more appropriate in municipal settings, where very low arsenic levels are required. Other techniques exist for arsenic removal, but are less well documented. When arsenic-rich water also contains high levels of dissolved iron, iron removal will also remove much of the arsenic. Introduction of zero-valent iron filings in three-pitcher filters to treat water in the home is showing great promise. Many new materials are being tested for arsenic removal, from lowtech iron-coated sand and greensand to specially engineered synthetic resins. In all cases, technologies should meet several basic technical criteria. The biggest challenges ahead lie however in applying the technologies described in poor, rural settings, and in enabling those communities to choose safe sources of water for drinking and cooking.

In areas where the drinking water supply contains unsafe levels of arsenic, the immediate concern is finding a safe source of drinking water. There are two main options: finding a new safe source, and removing arsenic from the contaminated source. In either case, the drinking water supplied must be free from harmful levels of arsenic, but also from bacteriological contamination, and other chemical contaminants. This paper reviews available technologies for arsenic removal.

When arsenic contamination is identified, the immediate priority must be to find a safe alternate source of drinking and cooking water for affected communities. Alternate sources must be not only arsenic-free, but also microbiologically safe—it would be a serious mistake to revert back to unsafe use of surface water sources. In some cases, there may be no one technology that can provide communities with a sustainable, continuous, affordable, safe water supply.

If a year-round safe water source is not currently available, it may be necessary as a short-term solution to use one source during wet seasons (e.g. groundwater, rainwater) and another during dry seasons (e.g. removing arsenic from contaminated water). If a completely satisfactory, arsenic-free water source cannot be established, the short-term goal should be to reduce arsenic levels in drinking water as much as possible, as quickly as possible, even if regulatory standards cannot be immediately met.

It should be recalled that health effects of arsenic are dose-dependent, and a partial solution is better than no solution. The implementation of a temporary solution should not be used as a reason to delay design and implementation of a long-term plan. In all cases, technologies should meet several basic technical criteria. Water supply options must first of all be able to produce water of the required quality, both chemical and bacteriological. Systems should also be able to supply water in adequate quantity, throughout different seasons. Technologies should be robust. It is important that operational safety be ensured. Finally, technologies should not have an undue adverse effect on the environment.

Technologies meeting these technical criteria can be evaluated under several socioeconomic criteria. First, the systems must be economically feasible. Introduction of new technologies requires institutional capacity including production and delivery of materials, training, and quality control monitoring. New options must be convenient, or people will not use them. Gender impacts should be considered, so that the workload of women and girls is not unduly increased.

New technologies require behavioural change on the part of the user, thus communication interventions should be considered. Finally, technologies must be socially acceptable to community members in order to be successful as a long-term safe water supply option.

TECHNOLOGIES FOR ARSENIC REMOVAL

In some areas, arsenic-contaminated water will be abundant, and arsenic-free sources scarce or polluted with other compounds. In these areas it may be most efficient to remove arsenic from the contaminated water, at least as a short term measure. Many technologies have been developed for the removal of arsenic. Most of the documented experience has been with large municipal treatment plants, but some of the same technologies can be applied at community or household levels. All of the technologies for arsenic removal rely on a few basic chemical processes, which are summarized below:

- *Oxidation/reduction*: reactions that reduce (add electrons to) or oxidize (remove electrons from) chemicals, altering their chemical form. These reactions do not remove arsenic from solution, but are often used to optimize other processes.
- *Precipitation*: Causing dissolved arsenic to form a low-solubility solid mineral, such as calcium arsenate. This solid can then be removed through sedimentation and filtration. When coagulants are added and form flocs, other dissolved compounds such as arsenic can become insoluble and form solids, this is known as coprecipitation. The solids formed may remain suspended, and require removal through solid/liquid separation processes, typically coagulation and filtration.
- *Adsorption and ion exchange*: various solid materials, including iron and aluminum hydroxide flocs, have a strong affinity for dissolved arsenic. Arsenic is strongly attracted to sorption sites on the surfaces of these solids, and is effectively removed from solution. Ion exchange can be considered as a special form of adsorption, though it is often considered separately. Ion exchange involves the reversible displacement of an ion adsorbed onto a solid surface by a dissolved ion. Other forms of adsorption involve stronger bonds, and are less easily reversed.
- *Solid/liquid separation*: precipitation, co-precipitation, adsorption,

and ion exchange all transfer the contaminant from the dissolved to a solid phase. In some cases the solid is large and fixed (e.g. grains of ion exchange resin), and no solid/liquid separation is required. If the solids are formed in situ (through precipitation or coagulation) they must be separated from the water. Gravity settling (also called sedimentation) can accomplish some of this, but filtration is more effective. Most commonly, sand filters are used for this purpose.

- *Physical exclusion*: some synthetic membranes are permeable to certain dissolved compounds but exclude others. These membranes can act as a molecular filter to remove dissolved arsenic, along with many other dissolved and particulate compounds.
- *Biological removal processes*: bacteria can play an important role in catalyzing many of the above processes. Relatively little is known about the potential for biological removal of arsenic from water.
- Boiling does not remove arsenic from water. Most of the established technologies for arsenic removal make use of several of these processes, either at the same time or in sequence. All of the removal technologies have the added benefit of removing other undesirable compounds along with arsenic—depending on the technology, bacteria, turbidity, colour, odor, hardness, phosphate, fluoride, nitrate, iron, manganese, and other metals can be removed.

Historically, the most common technologies for arsenic removal have been coagulation with metal salts, lime softening, and iron/manganese removal. Since the WHO Guideline Value for arsenic in drinking water was lowered from 50 to 10 μg/L in 1993, several countries have lowered their drinking water standards, in some cases to 10 μg/L.

In January 2001, the USEPA lowered the U.S. drinking water standard from 50 to 10 μg/L(only to postpone/reconsider this decision a few months later). Coagulation processes are sometimes unable to efficiently remove arsenic to these low levels. As a result, various alternate technologies have been developed or adapted that are capable of removing arsenic to trace levels. These advanced treatment options include ion exchange, activated alumina, and membrane methods such as reverse osmosis and nanofiltration.

While these technologies have all been shown to be effective in lab or pilot studies, there is still relatively little experience with full-scale treatment. In addition, a number of novel removal technologies are under development, some of which show great promise. The main arsenic removal technologies are presented below, along with a brief description of how removal efficiency is affected by arsenic concentration and speciation, pH, and the presence of other dissolved constituents.

OXIDATION

Most arsenic removal technologies are most effective at removing the pentavalent form of arsenic (arsenate), since the trivalent form (arsenite) is predominantly non-charged below pH 9.2 Therefore, many treatment systems include an oxidation step to convert arsenite to arsenate. Oxidation alone does not remove arsenic from solution, and must be coupled with a removal process such as coagulation, adsorption or ion exchange. Arsenite can be directly oxidized by a number of other chemicals, including gaseous chlorine, hypochlorite, ozone, permanganate, hydrogen peroxide, and Fenton's reagent (H_2O_2/Fe^{2+}).

Some solids such as manganese oxides can also oxidize arsenic. Ultraviolet radiation can catalyze the oxidization of arsenite in the presence of other oxidants, such as oxygen. Direct UV oxidation of arsenite is slow, but may be catalyzed by the presence of sulfite, ferric iron or citrate.

Chlorine is a rapid and effective oxidant, but may lead to reactions with organic matter, producing toxic trihalomethanes as a by-product. Chlorine is widely available globally, though if improperly stored it can lose its potency rapidly. In Europe, and increasingly in the USA, ozone is being used as an oxidant. In developing countries, ozone has not been widely used. An ozone dose of 2 mg/L, contacted with the water for 1 minute prior to filtration, has been shown to be effective in oxidizing iron and manganese, at the same time removing arsenic and other metals to below detection limits.

At a similar ozone dose, arsenite was shown to have a half-life of approximately 4 minutes. Ozone is also a potent disinfectant, but unlike chlorine, does not impart a lasting residual to treated water. Permanganate effectively oxidizes arsenite, along with Fe(II) and Mn(II). It is a poor disinfectant, though it can produce a bacteriostatic effect.

Potassium permanganate ($KMnO_4$) is widely available in developing countries, where it is used as a topical antibiotic for minor cuts. It is relatively stable with a long shelf life. Residual manganese in treated water should not exceed the WHO guideline of 0.5 mg/L. Hydrogen peroxide may be an effective oxidant if the raw water contains high levels of dissolved iron, which often occur in conjunction with arsenic contamination.

COAGULATION AND FILTRATION

The most heavily documented treatment methods for arsenic removal involve coagulation and filtration, either using metal salts or lime softening. This treatment can effectively remove many suspended and dissolved constituents from water besides arsenic, notably turbidity, iron, manganese, phosphate and fluoride. Significant reductions are also possible in odor, colour, and potential for trihalomethane formation.

Thus coagulation and filtration to remove arsenic will improve other water quality parameters, resulting in ancillary health and esthetic benefits. However, the optimal conditions vary for removal of different constituents, and coagulation to remove arsenic may not be optimal for removal of other compounds, notably phosphate and fluoride. Arsenic removal with metal salts has been shown since at least 1934. The most commonly used metal salts are aluminum salts such as alum, and ferric salts such as ferric chloride or ferric sulfate.

Ferrous sulfate has also been used, but is less effective. Excellent arsenic removal is possible with either ferric or aluminum salts, with laboratories reporting over 99% removal under optimal conditions, and residual arsenic concentrations of less than 1 μg/L. Fullscale plants typically report a somewhat lower efficiency, from 50% to over 90% removal. During coagulation and filtration, arsenic is removed through three main mechanisms:

- *Precipitation*: the formation of the insoluble compounds $Al(AsO_4)$ or $Fe(AsO_4)$
- *Coprecipitation*: the incorporation of soluble arsenic species into a growing metal hydroxide phase
- *Adsorption*: the electrostatic binding of soluble arsenic to the external surfaces of the insoluble metal hydroxide.

All three of these mechanisms can independently contribute towards contaminant removal. In the case of arsenic removal, direct precipitation has not been shown to play an important role. However, coprecipitation and adsorption are both active arsenic removal mechanisms. Numerous studies have shown that filtration is an important step to ensure efficient arsenic removal.

After coagulation and simple sedimentation, HAO and HFO – along with their sorbed arsenic load – can remain suspended in colloidal form. Hering and others showed that coagulation and sedimentation without filtration achieved arsenate removal efficiencies of 30%; after filtration through a 1.0 micron filter, efficiency was improved to over 96%. Only marginal improvements were made by reducing the filter size to 0.1 micron. In field applications, some plants improve arsenic removal with two-stage filtration.

ION-EXCHANGE RESINS

Synthetic ion exchange resins are widely used in water treatment to remove many undesirable dissolved solids, most commonly hardness, from water. These resins are based on a cross-linked polymer skeleton, called the 'matrix'. Most commonly, this matrix is composed of polystyrene cross-linked with divinylbenzene.

Charged functional groups are attached to the matrix through covalent bonding, and fall into four groups:

- Strongly acidic (e.g. sulfonate, $-SO_3^-$)
- Weakly acidic (e.g. carboxylate, $-COO^-$)
- Strongly basic [e.g. quaternary amine, $-N^+(CH_3)_3$]
- Weakly basic [e.g. tertiary amine, $-N(CH_3)_2$]

The acidic resins are negatively charged, and can be loaded with cations (e.g. Na^+), which are easily displaced by other cations during water treatment. This type of cation exchange is most commonly applied to soften hard waters. Conversely, strongly basic resins can be pretreated with anions, such as Cl-, and used to remove a wide range of negatively charged species. Clifford gives the following relative affinities of some common anions for a type 1 strong-base anion resins:

$CrO_4^{2-} >> SeO_4^{2-} >> SO_4^{2-} >> HSIO_4^- > HO_3^- > Br^- > HAsO_4^{2-} SeO_3^{2-} > HSO_3^{3-} > NO_2^- > Cl^-$

Different resins will have differing selectivity sequences, and resins have been developed specifically to optimize removal of sulfate, nitrate, and organic matter. Various strong-base anion exchange resins are commercially available which can effectively remove arsenate from solution, producing effluent with less than 1 ìg/L arsenic. Arsenite, being uncharged, is not removed. Analysts have taken advantage of this specificity to develop procedures for analytical differentiation of arsenite and arsenate.

Therefore, unless arsenic is present exclusively as arsenate, an oxidation step will be a necessary precursor to arsenic removal. Conventional sulfate-selective resins are particularly suited for arsenate removal. Nitrate-selective resins also remove arsenic, but arsenic breakthrough occurs earlier. Most commonly, resins are pretreated with hydrochloric acid, to establish chloride ions at the surface, which are easily displaced by arsenic, though the resin can be primed with other anions such as bromide or acetate.

Packed beds are commonly designed to have an Empty Bed Contact Time (EBCT) of 1.5 to 3 minutes. Arsenate removal is relatively independent of pH and influent concentration. On the other hand, competing anions, especially sulfate, have a strong effect. The number of bed volumes that can be treated before arsenic breakthrough (defined as 10% of the influent concentration) can be roughly estimated with two simple formulas:

For $[SO_4^{2-}] < 120$ mg/L : Bed Volumes = – 606 *ln$[SO_4^{2-}]$ + 3,150

For $[SO_4^{2-}] > 120$ mg/L : Bed Volumes = – 200 *ln$[SO_4^{2-}]$ + 1,250

where $[SO_4^{2-}]$ is the initial sulfate concentration in mg/L. In low-sulfate waters, ion exchange resin can easily remove over 95% of arsenate, and treat from several hundreds to over a thousand bed volumes before arsenic breakthrough occurs. Accordingly, the USEPA recommends that ion exchange resins not be used in waters with >120 mg/L sulfate or >500 mg/L TDS, and will be most effective in waters with even lower sulfate levels (<25 mg/L).

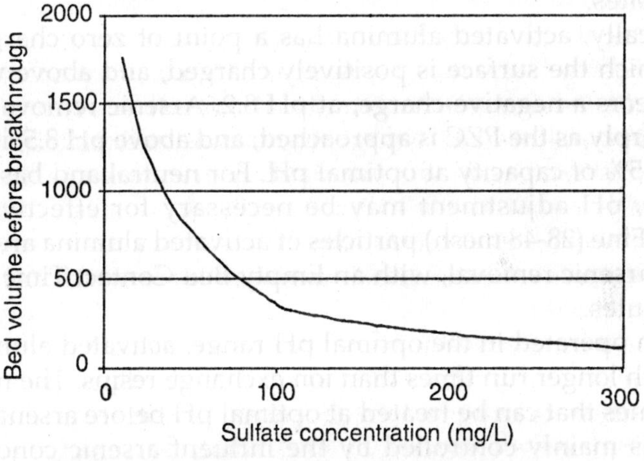

Fig. Bed Volumes Treated with Ion Exchange Resin

Ion exchange capacity, analogous to the adsorption capacity discussed in the previous section, is a measure of the number of exchange sites, and is usually measured in milliequivalents (meq) per mL wet volume, including pore spaces. The operating capacity measures actual performance of resins under environmental conditions, and is always less than the advertised exchange capacity, due to incomplete regeneration and contaminant leakage.

Activated Alumina

Activated alumina is a granulated form of aluminum oxide (Al_2O_3) with very high internal surface area, in the range of 200-300 m^2/g. This high surface area gives the material a very large number of sites where sorption can occur, and activated alumina has been widely used for removal of fluoride. In the early 1970s Bellack accidentally discovered that activated alumina could remove arsenic from water.

The mechanisms of arsenic removal are similar to those of a weak base ion exchange resin, and are often collectively referred to as 'adsorption', though ligand exchange and chemisorption are technically more appropriate terms. The kinetics of arsenic removal onto the alumina surface are slower than those of ion exchange resins, and some arsenic leakage is often noted in activated alumina systems.

Arsenic removal efficiency is excellent (typically > 95%), for both arsenate and arsenate, but arsenic capacity varies significantly, and is controlled primarily by pH and influent arsenic concentration and speciation. Arsenate removal capacity is best in the narrow range from pH 5.5 to 6.0, where the alumina surfaces are protonated, but acid anions are not yet concentrated enough to compete with arsenic for sorption sites.

Typically, activated alumina has a point of zero charge (PZC), below which the surface is positively charged, and above which the surface bears a negative charge, at pH 8.2. Arsenic removal capacity drops sharply as the PZC is approached, and above pH 8.5, is reduced to only 2-5% of capacity at optimal pH. For neutral and basic waters, therefore, pH adjustment may be necessary for effective arsenic removal. Fine (28-48 mesh) particles of activated alumina are typically used for arsenic removal, with an Empty Bed Contact Time of five to eight minutes.

When operated in the optimal pH range, activated alumina beds have much longer run times than ion exchange resins. The number of bed volumes that can be treated at optimal pH before arsenate breaks through is mainly controlled by the influent arsenic concentration Frank and Clifford reported an arsenate capacity (at pH 6) of about 1.6 g/L of activated alumina, consistent with an earlier reported capacity of 4 mg/g, assuming a bulk density of 0.5 kg/L.

Fox reported a somewhat lower capacity of 1 mg/g, but this is likely due to the elevated pH (7.4- 8.0) of the influent water. The sorption sites on the activated alumina surface are also attractive to a number of anions other than arsenate: Clifford reports the selectivity sequence of activated alumina in the pH range of 5.5 to 8.5 as:

$$OH^- > H_2AsO_4^- > Si(OH)_3O^- > HSeO_3^- > F^- > SO_4^{2-} > CrO_4^{2-} >> HCO_3^- > Cl^- > NO_3^- > Br^- > I^-$$

Trussell and others reported a similar selectivity sequence, but included phosphate as the second most preferred anion, after hydroxyl, and placed fluoride above arsenate in the sequence. Because of activated alumina's strong selectivity for arsenate, competing anions pose less of a problem than with ion exchange resins. Sulfate, and to a lesser extent, chloride, have been shown to reduce capacity, but the competition effect is not as dramatic as with ion exchange resins. Phosphate and fluoride are also sorbed onto activated alumina, producing improvements in drinking water quality, but at the same time reducing arsenic removal potential.

Activated alumina can be regenerated by flushing with a solution of 4% sodium hydroxide, which displaces arsenic from the alumina surface, followed by flushing with acid, to re-establish a positive charge on the grain surfaces. Regeneration is more difficult, and less complete (generally 50-80%) than with ion exchange resins.

The advantages of activated alumina are that simple removal systems can be developed at community or household levels that require no chemical addition. Since activated alumina can treat thousands of bed volumes before breakthrough, filters could be operated for months before the media need to be changed or regenerated. Activated alumina will also remove selenite, fluoride, sulfate, and chromate.

Disadvantages include the possibility that the media will be fouled or clogged by precipitated iron, the relatively narrow pH range for optimal operation, and the relative difficulty of regeneration. Also, compared with ion exchange resins, a significantly longer Empty Bed Contact Time is required.

Membrane Methods

Synthetic membranes are available which are selectively permeable: the structure of the membrane is such that some molecules can pass through, while others are excluded, or rejected. Membrane filtration has the advantage of removing many contaminants from water, including bacteria, salts, and various heavy metals. Two classes of membrane filtration can be considered: low-pressure membranes, such as microfiltration and ultrafiltration; and high-pressure membranes such as nanofiltration and reverse osmosis.

Low-pressure membranes have larger nominal pore sizes, and are operated at pressures of 10-30 psi. The tighter high-pressure membranes are typically operated at pressures from 75 to 250 psi, or even higher.

From Figure, it is clear that reverse osmosis (RO) and nanofiltration (NF) membranes have pore sizes appropriate for removal of dissolved arsenic, which is in the 'metal ion' size range. Both RO and NF membranes are most often operated in lateral configurations, in which only a small amount of the raw water (10-15%) passes through the membrane as permeate. In household systems, where only a small amount of treated water is required for cooking and drinking, this low recovery rate may be acceptable. Municipal systems achieve higher recovery rates (80 to over 90%) by using multiple membrane units in series.

In recent years, a new generation of RO and NF membranes have been developed that are less expensive and operate at lower pressures, yet allow improved flux and are capable of efficient rejection of both arsenate and arsenite.

Waypa and others have showed that some of the new membranes, operated at pressures ranging from 40-400 psi, were able to reject from 96-99% of both arsenate and arsenite in spiked natural waters. The authors attribute this rejection of arsenite to the relatively large molecular weight of both arsenate and arsenic, rather than charge repulsion. At these high arsenic rejection rates, membrane filtration can result in extremely low arsenic levels in treated water.

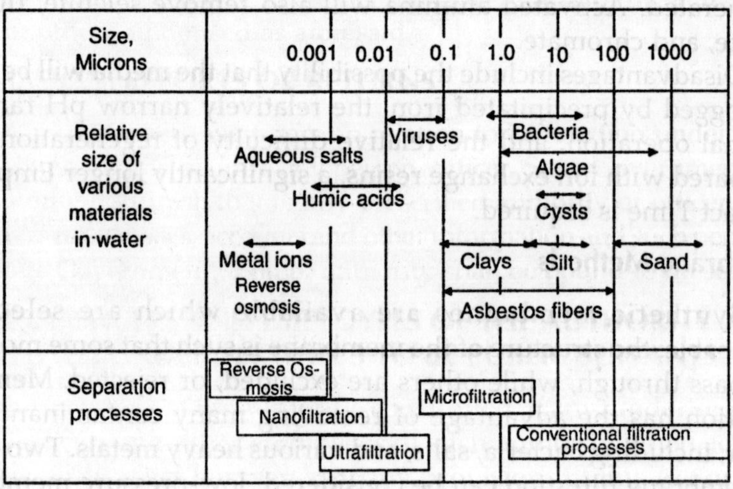

Fig. Pore Size of Various Membranes, and Size of Materials Subject to Filtration.

Arsenic removal was found to be independent of pH and the presence of cooccurring solutes, but was somewhat improved at lower temperatures. Interestingly, the NF membrane tested performed comparably to the RO membranes, even though the operating pressure was much lower (40-120 psi, compared to 200-400 psi). Membrane filtration requires a relatively high-quality influent water. Membranes can be fouled by colloidal matter in the raw water, particularly organic matter. Iron and manganese can also lead to scaling and membrane fouling. To prevent fouling, reverse osmosis filters are almost always preceded by a filtration step. Membrane filtration has the advantage of lowering the concentrations of many other components in addition to arsenic.

Even ultrafiltration (UF) membranes are able to remove over 99.9% of bacteria, Giardia and viruses. Also, the membrane itself does not accumulate arsenic, so disposal of used membranes would be simple. Operation and maintenance requirements are minimal: no chemicals need be added, and maintenance would consist of ensuring a reasonably constant pressure, and periodically wiping the membrane clean.

The main disadvantages are low water recovery rates (typically only 10-20% of the raw water passes through the membrane), the need to operate at high pressures, relatively high capital and operating costs, and the risk of membrane fouling. Also, particularly with RO, the treated water has very low levels of dissolved solids, and can be very corrosive, and deficient in minerals which can be important micronutrients for humans.

EMERGING TECHNOLOGIES

In recent years, a tremendous amount of research has been conducted to identify novel technologies for arsenic removal, particularly low-cost, low-tech systems that can be applied in rural areas. Most of these technologies rely on oxidation of arsenite, followed by filtration through some sort of porous material, where arsenic is removed through adsorption and coprecipitation. Many of these systems make use of iron compounds, which have a very strong affinity for arsenic. A brief review of some of the most documented technologies is given below.

Fe-Mn Oxidation

Conventional iron and manganese removal can result in significant arsenic removal, through coprecipitation and sorption onto ferric or manganic hydroxides. The mechanisms involved are the same as in coagulation and filtration. Most low-cost technologies for arsenic and manganese removal rely on aeration and filtration through porous media such as sand and gravel.

Any technology that effectively removes iron and manganese could be evaluated to see if arsenic is also removed effectively. In this respect arsenic removal is more convenient than that of fluoride, which does not undergo oxidation, and is not removed by coprecipitation with iron. In Bangladesh and West Bengal, elevated arsenic concentrations are often associated with high iron and manganese levels. One survey in Bangladesh found that over 80% of arsenic-affected tubewells (>50 ìg/L) also contained iron levels of 2 mg/L or more. However, iron alone is not a good indicator of arsenic: 30% of the wells with safe levels of arsenic also had 2 mg/L iron or more.

Because of the link between arsenic and iron levels, and the affinity of arsenic for iron hydroxides, there have been calls for a simple solution to arsenic contamination: simple storage of pumped water to allow iron to settle out, scavenging arsenic in the process. While this is an appealing idea, successful application of this type of 'passive Fe-Mn oxidation' is not simple, for several reasons:

- Iron removal is not always easily accomplished. Some waters contain iron in a form that is slow to oxidize, or may be complexed with organic material that impedes oxidation and filtration. Precipitation may not occur if alkalinity is low.
- Without a filtration step, much of the iron can remain suspended as colloidal matter, even after oxidation.
- Arsenite is not as strongly bound to iron as arsenate, if the waters contain mostly arsenite arsenic removal will be less efficient.
- When water is stored in household containers, there is a high risk of bacterial contamination.

When considering passive Fe-Mn oxidation, particularly at the household level, careful pilot studies should be made using the local waters and local storage conditions, in order to assess the effectiveness of this technique, and the possibility of pathogenic contamination. It should be noted that chlorine addition would improve oxidation of both iron and arsenic, and would provide protection against bacterial growth. However, as discussed above, chlorination at the household level involves difficulties in ensuring the correct dose, and the potency of the chlorine agent. With support from the Dutch Government, the Department of Public Health Engineering of Bangladesh has constructed three arsenic removal plants in small municipalities. These plants are basically iron removal plants, and add no chemicals, but pump groundwater over a series of cascades to aerate the water.

Filtration then removes the resulting iron and arsenic precipitate, and the water is chlorinated and stored in an elevated tank for distribution. Water stored in the tank is periodically used to backwash the filters. The waste water is stored in sludge ponds, and sludge is removed once or twice annually. Arsenic removal efficiency varies considerably, and seems to improve with higher iron levels:

Table. Arsenic Removal in Three 18-DTP Plants

Municipality	Influent iron (mg/L)	Iron removal efficiency	Influent arsenic (μg/L)	Arsenic removal efficiency (%)
Satkira Razzak	3.4	95	57.0	51
Satkira Polash	5.8	95	67.5	67
Manikganj	7.6	99	84.8	72

Although removal rates are not very high, in all cases it is effective enough to bring waters into compliance with the Bangladesh drinking water standard of 50 μg/L. Plant managers experimented with addition of coagulants (4 mg/L $FeCl_3$) and oxidants (0.9 mg/L bleaching powder), but found that arsenic removal efficiency was not significantly improved.

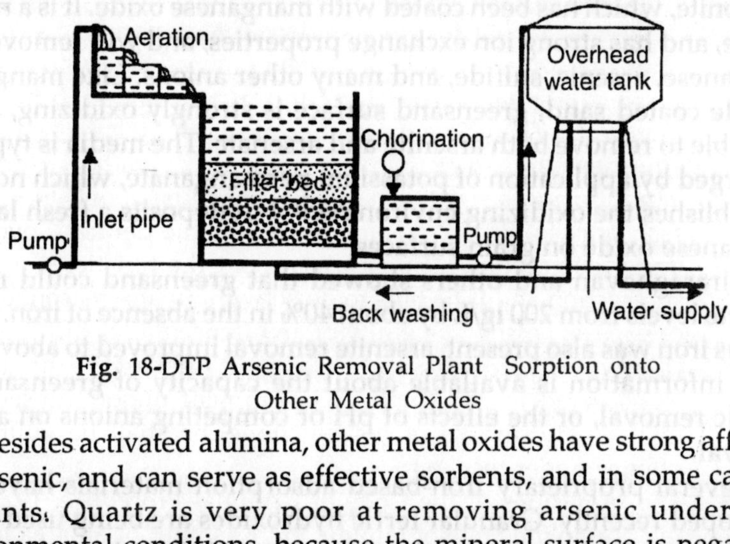

Fig. 18-DTP Arsenic Removal Plant Sorption onto Other Metal Oxides

Besides activated alumina, other metal oxides have strong affinities for arsenic, and can serve as effective sorbents, and in some cases as oxidants. Quartz is very poor at removing arsenic under most environmental conditions, because the mineral surface is negatively charged above a pH of 2. However, quartz sand, or indeed any other granular media, can be made highly sorptive by coating the grains with metal oxides.

In recent years many researchers have used this principle to develop low-cost arsenic removal methods using locally available materials. Vaishya showed that sand from the Ganges river, which presumably is rich in iron coatings, could remove arsenite from solution, with a reported capacity of 0.024 mg/g. Removal was found to be pH-dependent, and best from pH 7-9.

Joshi and Chaudhuri showed that iron oxide coated sand (IOCS) is able to remove both arsenite and arsenate. A simple fixed bed unit was able to treat about 160-190 bed volumes of water containing 1000 ìg/L arsenite and 150-165 bed volumes of water with 1000 ìg/L arsenate. Flushing with 0.2 N sodium hydroxide regenerates the media. The authors propose that this media would be very useful for domestic arsenic removal units.

A similar coated sand material can be prepared using manganese dioxide instead of iron. Since MnO_2 is a good oxidant, this material can remove arsenite as well as arsenate. In fact, the treated sand was able to remove 80% of a 1 mg/L solution of arsenite within two hours, but slightly less than 70% of an equivalent solution of arsenate. A prototype household unit was developed, which could treat about 150 bed volumes of 1 mg/L arsenic (half arsenite and half arsenate) before breakthrough.

Greensand is a granular material composed of the mineral glauconite, which has been coated with manganese oxide. It is a natural zeolite, and has strong ion exchange properties, and will remove iron, manganese, arsenic, sulfide, and many other anions. Like manganese dioxide coated sand, greensand surface is strongly oxidizing, and is thus able to remove both arsenite and arsenate. The media is typically recharged by application of potassium permanganate, which not only reestablishes the oxidizing environment but deposits a fresh layer of manganese oxide on grain surfaces.

Viraraghavan and others showed that greensand could reduce arsenite levels from 200 ìg/L by about 40% in the absence of iron. When ferrous iron was also present, arsenite removal improved to above 80%. Little information is available about the capacity of greensand for arsenic removal, or the effects of pH or competing anions on arsenic removal.

Several proprietary iron-based adsorption materials have been developed recently. Granular ferric hydroxides are being used in full scale systems in Germany, and similar materials have been developed in Canada and the United States. These materials generally have high removal efficiency and capacity.

Sorption onto Reduced Metals

Most of the above processes rely on arsenate adsorption onto surfaces of metal oxides. However, arsenic also has a strong affinity to reduced metal surfaces, such as sulfides. A few researchers have taken advantage of this property to remove arsenic through reduction and sorption. Lackovic and others have demonstrated that zero-valent iron filings can be used either in situ or ex situ to reduce arsenate, and produce ferrous iron. The ferrous ions precipitate out with sulfide, which is also added to the system. Arsenite is removed either through coprecipitation or adsorption onto pyrite. This system is promising for use in rural areas, because of the low cost of materials, and the simple operation.

However, treated water is very high in ferrous iron, and must undergo iron removal treatment before distribution or consumption. A similar system using zero-valent iron to treat water stored in individual homes was tested in Bangladesh and West Bengal (the so-called: three kolshi filter). Arsenic removal was approximately 95% for highly contaminated waters, containing 2000 ìg/L arsenic in the presence of sulfate at pH 7. Removal is rapid, but if batches are left for too long, dissolved iron concentrations become unacceptably high.

Ion Exchange and Adsorption

Ion exchange resins developed for removal of anions such as sulfate and nitrate have proven to be reasonably effective at removing arsenic, as discussed above. As materials engineering becomes more advanced, researchers are increasingly able to design novel ion-exchange materials with surficial properties that are particularly specific to arsenate. In particular, several researchers have found that copper-doped materials have a strong, specific affinity for arsenate.

Fryxell and others have developed a novel mesoporous silica sorbent which makes use of Cu(II)-based functional groups. This material has a higher ion exchange capacity (75 mg As/g) than conventional resins, and shows a stronger affinity for arsenate and chromate than for sulfate or nitrate. Therefore, unlike the conventional resins, these materials will not release chromatographic peaks of arsenic when exposed to high levels of sulfate.

While conventional synthetic ion exchange resins are the most commonly used media in ion exchange, costs are relatively high ($USD 2-5 per litre of wet resin). A variety of naturally occurring materials also have high ion exchange capacities, sometimes after chemical pretreatment. Many of these materials are not pure ion exchangers: some arsenic removal is through less reversible chemisorption. Especially in developing countries, researchers have been evaluating the potential of these materials for use as low-cost arsenic removal systems.

Zeolites are naturally occurring minerals with a crystalline structure characterized by large internal pore spaces. Accordingly, they have very large surface areas, and ion exchange capacities: zeolites were used extensively for water softening, before the development of synthetic resins with faster exchange rates, higher capacity, and longer life. A few arsenic removal studies have been conducted with zeolites. Natural zeolite minerals such as clinoptilolite and chabazite have a strong affinity for both arsenite and arsenate. A chabazite filter was able to remove 1000 ìg/L arsenate from over 235 bed volumes before arsenic was detected in the effluent. Adsorption of arsenate onto natural zeolites can be improved by organically modifying the zeolite structure.

Chitosan and chitin are natural polyaminosaccharides occurring in crustacean shells, that have good ion exchange properties. Shellfish wastes containing chitosan have been used to remove arsenic from water contaminated by mining wastes. Elson and others investigated a mixture of chitosan and chitin, and found a relatively low arsenic removal capacity of about mg As/g.

In Situ Arsenic Immobilization

When arsenic is mobilized in groundwater under reducing conditions, it is possible to immobilize the arsenic by creating oxidized conditions in the subsurface. In Germany, in order to remediate an aquifer containing higharsenite, high ferrous iron, low-pH groundwater, Matthess injected 29 tons of potassium permanganate directly into 17 contaminated wells, oxidizing arsenite, which coprecipitated out with ferric oxides. Mean arsenic concentrations were reduced by over 99%, from 13,600 to 60 ìg/L.

More recently, atmospheric oxygen was used to reduce arsenic concentrations in situ from approximately 20 to 5 ìg/L, while iron and manganese levels were also lowered. Under reducing conditions, and in the presence of sulfur, arsenic can precipitate out of solution and form relatively insoluble arsenic sulfides In situ immobilization has the great advantage of not producing any wastes that must be disposed of. However, experience is limited, and the technique should be considered with caution.

Oxidants are by definition reactive compounds, and may have unforeseen effects on subsurface ecological systems, as well as on the water chemistry. Care must also be taken to avoid contaminating the subsurface by introducing microbes from the surface. Also, at some point pore spaces can become clogged with precipitates, particularly if dissolved iron and manganese levels are high in the untreated water.

The Table summarizes some of the key technologies for arsenic removal, with special reference to experiences gained from field level application. Research needs are also identified. Arsenic removal efficiency will vary according to many site-specific chemical, geographic, and economic conditions, so actual applications may vary from the generalizations listed below. Because of the many factors that can affect arsenic removal efficiency (including arsenic concentration, speciation, pH and co-occurring solutes), any technology should be tested using the actual water to be treated, before implementation of arsenic removal systems at the field scale.

Table. Summary of Technologies for Arsenic Removal

Technology	Removal Efficiency		Institutional Experience and Issues
	As (III)	As (V)	
Coagulation with iron salts	++	++	Well proven at central level, piloted at community and household levels. Phosphate and silicate may reduce arsenic removal rates. Generates arsenic-rich sludge. Relatively inexpensive.
Coagulation with alum	–	++	Proven at central level, piloted at household levels. Phosphate and silicate may reduce arsenic removal rates. Optimal over a relatively narrow pH range. Generates arsenic-rich sludge. Relatively inexpensive.
Lime softening	+	+++	Proven effective in laboratories and at pilot scale. Efficiency of this chemical process should be largely independent of scale. Chiefly seen in central systems in conjunction with water softening. Disadvantages include extreme pH and large volume of waste generated. Relatively inexpensive, but more expensive than coagulation with iron salts or alum because of larger does required, and waste handlings.
Ion exchange	–	+++	Pilot scale in central and household systems, mostly in industrialized countries. Interference from sulfate and TDS. High adsorption capacity, but long-term performance of regenerated media needs documentation. Wates rich in iron and manganese may require pre-treatment toprevent media clogging. Moderately expensive. Regeneration produces arsenic-rich brine.
Activated alumina	+/++	+++	Pilot scale in community and household systems, in industrialized and developing countries, Arsenite removal is poorly understood, but capacity is much less than for arsenate. Regeneration requires strong acid and base, andproduces arsenic-rich waste. Long-term performance of regenerated media needs documentation. Waters rich in iron and manganese

			may require pre-treatment to prevent media clogging. Moderately expensive.
Membrane	–/+++	+++	Shown effective in laboratoey studies in industrialized countries. Research neededonremoval of arsenite, and efficiency at high recovery rates, especially with low-pressure membranes. Pretreatment usually required. Relatively expensive, especially if operated at high pressures.
Fe-Me oxidation	?	+/++ +++	Small-scale application in central systems, limited studies in community and household levels. More research neededon which hydrochemical conditionsare conducive for good arsenic removal. Inexpensive.
Porous media sorbents (iron oxide coated sand, greensand, etc.)	+/++	++/+++	Shown effective in laboratoey studies in industrialized and developing countries. Need to be evaluated under different environmental conditions, and in field settings. Simple media re inexpensive, advanced media can be relatively expensive.

Key: +++ Consistently > 90% removal
++ Generally 60-90% removal
+ Generally 30-60% removal
– < 30% removal
? Insufficient information

REMOVAL OF ARSENIC FROM DRINKING WATER AND SOIL BIOREMEDIATION

Various countries in America have reported the existence of people suffering chronic exposure to levels of arsenic in drinking water that are higher than those envisaged in international standards. This is the case in Canada, the United States, Chile, Peru, Bolivia, Mexico, El Salvador and Nicaragua. Some of these countries have solved, either fully or in part, the problem of technology availability, depending on whether the affected population lives in rural or urban areas.

There are around 14 methods of removing arsenic from water, with efficiencies that range from 70 to 99%. Coagulation-flocculation and lime softening are the methods most used in large systems, and not only to remove arsenic. Small systems can use ion exchange, activated alumina, reverse osmosis, nanofiltration and reverse electrodialysis. New technologies include sand covered with iron oxides, granular ferric hydroxide, iron packets, iron modified by sulfur, zeolite filtration, iron addition with direct filtration and conventional removal of iron and manganese. In Latin America studies have concentrated on the use of chemical coagulation with aluminum sulfate, hydrated lime and sodium polyelectrolyte, and have achieved arsenic levels of 0.12-0.15 mg/l. Direct coagulation over filters and coagulation-flocculation have reached values below 0.05 mg/l. Removal through adsorption uses hematites and materials with a high iron content and positively charged surfaces (natural green clay, activated clays, natural and activated zeolite and bone charcoal).

In Latin America, Chile is the most experienced country in treating water for urban areas since it has four plants for removing arsenic from the water supply (0.40 mg/l) which together treat 2000 l/s and produce drinking water with 0.040 mg As/l. Improving the system by adding reverse osmosis (post treatment) and desalination has also been evaluated. In Peru there is one plant for removing arsenic that treats the water with ferric chloride and sulfuric acid. CEPIS-BS/SDE/PAHO has developed and patented a product called ALUFLOC, which is a mixture of an oxidant, activated clays and a coagulant (aluminum sulfate or ferric chloride).

This is a simple, low-cost methodology which, at a household level, removes the natural arsenic present in groundwater that is consumed by the rural population. Up to 98% has been removed using $Al_2(SO)_3$ and $FeCl_3$ as coagulants. As far as soil remediation in contaminated areas is concerned, studies have been carried out on the ability of certain plants to absorb and concentrate toxic substances. The University of Florida has identified a fern which absorbs and hyper-accumulates arsenic from contaminated soil.

Techniques used to restore and stabilize arsenic in sludge, soil and industrial waste generally involve precipitation with lime and caustic soda. This is followed by sedimentation and/or filtration. In Mexico an insoluble arsenic compound has been produced which can also be used as a raw material in the manufacture of solid products for use in construction or disposed of in a landfill.

In addressing the problem of drinking water, the characteristics of the water source, water treatment, and methods for distribution and/or consumption must be taken into account, as well as the type of technology to be employed, which will depend on the characteristics of the location. Latin American countries have the experience and ability to develop technologies, but are limited by a lack of financial resources, facilities and, above all, state policies to facilitate and direct the development of such technologies which will lead to an effective solution to the problem or to satisfy existing needs.

The people who are most affected are scattered over rural areas, consume untreated water and are unaware of the risks to which they are exposed. Pilot studies must be carried out on an ongoing and sustained basis until a solution is found that can be recommended for implementation in national programs to remove arsenic from drinking water. Several countries in America have reported people suffering chronic exposure to arsenic in drinking water at higher than permitted concentrations.

This has occurred in Canada, the United States, Chile, Peru, Bolivia, Mexico and El Salvador. Some countries have solved the problem, either fully or partially, depending basically on whether the affected population lived in rural or urban areas. This document is a collection of the available information on the treatment of water containing arsenic, and aims to show the existing treatment alternatives. It is complemented by the document "The Presence of Arsenic in Drinking Water in Latin America and its Effect on Public Health" submitted to this Congress by the authoress.

WATER TREATMENT

In general, drinking water is treated to remove colour, turbidity and fecal microorganisms. This is achieved through a suitable combination of processes: coagulation-flocculation-sedimentation-filtration and disinfection. But when it is intended to remove chemicals like arsenic from water, more complex methods are required, such as those given in the chart:

Techniques used to remove arsenic:

- Oxide reduction
- Solid-liquid separaton:
  - Lime softening
  - Coagulation-adsorption-filtration
    - Coagulation-adsorption
    - Coagulation-Filtration
    - Using iron and aluminum salts
    - Presence of iron and manganese
    - Granular ferric hydroxide
    - Iron with diret diltration
    - Activated alumina
  - Iron exchange
- Physical separation
  - Reverse electrodialysis
  - Reverse osmosis and nanofiltration
- Biological processes

There are around 14 technologies for removing arsenic from water, with efficiencies which vary from 70% to 99%. The coagulation – flocculation and lime softening methods are the most used in large systems, and not exclusively to remove arsenic. Small systems can apply ion exchange, activated alumina, reverse osmosis, nanofiltration and reverse electrodialysis.

New technologies make use of sand covered with iron oxides, granular ferric hydroxide, iron packets, sulfur-modified iron, zeolite filtration, the addition of iron with direct filtration and conventional removal of iron and manganese.

In the majority of cases, the efficiency of the chosen process depends on the initial concentration, state of oxidation of the arsenic and the pH. The catalytic action of light and the use of bacteria and spores are also being studied. Some methods for removing arsenic are described below.

Oxidation /Reduction

This process oxidizes arsenite into arsenate to improve its removal in complementary processes. Chlorine, chlorine dioxide, ozone and potassium permanganate can be used. Catalytic oxidation of As^{+3} is possible in the presence of copper oxide, activated carbon and UV radiation. One of the problems with this process is the reaction time. Biological oxidation is also possible as is removal through the catalytic action of light.

Solid/Liquid Separation

The processes of precipitation, co-precipitation, adsorption and ion exchange can be used to transfer arsenic from the dissolved phase to the solid phase. In some cases the solid which provides the adsorption surface is large and fixed, e.g. ion exchange resins and additional separation is required.

Solids formed in situ (through precipitation or coagulation) must be separated from the water by sedimentation or filtration. Precipitation: The dissolved arsenic is transformed into a low-solubility solid and removed by sedimentation and filtration. For example, precipitation of calcium arsenate; certain dissolved compounds such as arsenic can also be co-precipitated during coagulation and flocculation, to then be removed by filtration.

Adsorption and ion exchange: Various solid materials, including iron and aluminum hydroxide flocs can adsorb arsenic through a surface adsorption mechanism, and thus be removed from the water. Ion exchange involves the reversible displacement of an ion attached to a solid surface by the As^{+5} and As^{+3} ions. This can be considered as a special form of adsorption although it is frequently considered separately.

Lime Softening

Softening with lime is a process similar to coagulation with metallic salts. Lime $Ca(OH)_2$ is hydrolyzed and reacts with the carbonic acid to form calcium carbonate, which acts as the adsorption agent in the removal of arsenic.

This process is typically used only with very hard water and with treatment at pH in the range of 10 to 12. This technique is not suitable for small systems because of its high cost.

The removal (jar test) of arsenic As^{+5} from water (river, well and other) having a concentration of 0.1 to 20 mg/l is 40-70% for a pH range of 9-10 and removal efficiency is increased when lime softening is followed by coagulation using iron. Lime softening when the pH range is 10.6-11.4 showed a high removal of As+5, up to 95% when the initial arsenic concentration in the water was 12 mg/l.

The main mechanism for removing arsenic through softening of water containing magnesium is adsorption of the arsenic in the magnesium hydroxide formed during softening. Arsenate removal is also excellent at pH > 11, and poorer at pH < than 10. Arsenic removal improves with the addition of iron.

But carbonate reduces the effect. Arsenic removal is reduced in the presence of traces of orthophosphate, especially at pH < 12. With this method the removal of As+3 is poor, with arsenite sorption densities an order of magnitude lower than for arsenate.

Coagulation-filtration-adsorption

Arsenic removal through coagulation can be combined with filtration and/or adsorption, therefore the best application conditions should be studied according to the characteristics of the water to be treated.

Coagulation-filtration

In water treatment plants, As^{+5} can be removed effectively by coagulation with aluminum or iron sulfate, and lime-soda softening. The coagulants mentioned are hydrolyzed to form hydroxides, on which the As^{+5} is adsorbed and co-precipitated with other metallic ions. According to the literature, natural water containing a large quantity of colloids requires high concentrations of coagulants to achieve the removal efficiencies shown in the following table:

Table. Efficiency of Coagulants at Removing Arsenic

Coagulant	Arsenate, As^{+5}		Arsenite As^{+3}	
	Removal (%)	pH	Removal (%)	pH
Ferric sulfate Fe_2 $(SO_4)^3$	100	< 9.0	20	< 9.0
Aluminum sulfate $Al_2(SO_4)_3$	90	< 7.0	50	< 7.0

Arsenic removal in conventional coagulation processes using aluminum and iron salts and soda softening depends on the initial concentration of this element, the pH of the water being treated and the type and dosage of the coagulant.

Coagulation - adsorption

Coagulation-adsorption with iron or aluminum salts is the most widely documented method of treatment, both for removing arsenic and dissolved and suspended compounds from water (arsenic, turbidity, iron, manganese, phosphate and fluorine). This method also produces significant reductions in odor, colour and trihalomethane precursors. Nevertheless the optimum conditions for arsenic removal will depend on the characteristics of the water and the treatment process conditions.

Because of the difficulty of removing As^{+3} by coagulation, it has to be oxidized to As^{+5}. In high and low pH ranges the efficiency of this method is reduced significantly. In order to ensure removal of arsenic, an important step is the use of filtration (e.g. sand filters). Aluminum sulfate is the coagulant most frequently used in water treatment because of its low cost and relative ease of handling, although other products such as ferrous and ferric sulfates, ferric chlorosulfate and ferric chloride, and alum and magnesium carbonate are also used.

These salts are hydrolyzed by the water into iron and aluminum hydroxides, which form gelatinous flocs that agglutinate, thus facilitating the process of separating dissolved and colloidal materials. The work done by Cheng *et al.* using coagulant dosages of 10, 20 and 30 mg/l at pH values of 7.0, 6.3 and 5.5 show that arsenic removal depends on the pH, the coagulant dosage and the turbidity of the untreated water. A high level of turbidity can affect the removal of As^{+5}, but this improves significantly when optimum doses of coagulants are used (to remove the turbidity) less than 20 mg/l. Of the two coagulants used, ferric chloride is more effective than aluminum sulfate (at similar dosages by weight). Removal of As^{+5} with aluminum sulfate depends on the pH (better results when the pH is less than 7), whilst arsenic removal using ferric chloride is less dependant on pH and improves at higher doses of coagulant. Studies of coagulation and adsorption for removing arsenic by coagulation with ferric chloride and adsorption in aqueous ferric oxide, have evaluated the influence of different variables (initial arsenic concentration and state of oxidation, adsorbent concentration or coagulant dosage, pH and presence of inorganic solutes).

It was observed that under similar conditions removal of As^{+5} is better than As^{+3}. For both forms of arsenic, removal depends on the dosage of coagulant and adsorbent concentration. Dosages greater than 5 mg/l of ferric chloride produced the best rates of removal. Coagulation is independent of the initial concentrations of arsenic, but with low initial concentrations of As^{+3} adsorption of this form of arsenic was better. The presence of sulfate ions and a pH lower than 7 make the removal of As^{+3} inefficient, whilst these conditions do not affect removal of As^{+5} significantly.

Removal of As^{+5} increases in the presence of calcium and when the pH is high. Removal of As^{+5} with ferric chloride at pH of 8-9 decreases when organic material is present. The efficiency of As^{+3} removal using ferric chloride is affected by the composition of the water, the presence of sulfate (at a pH of 4 to 5) and the natural organic material (at a pH of 4 to 9). During the removal of arsenic, oxidation of As^{+3} to As^{+5} is important as is the adsorption of the amorphous ferric hydroxide formed during coagulation. Removal of As^{+5} is equally effective using iron and aluminum salts (at the same molar concentration) at pH < 7.5.

Nevertheless, iron is more effective than aluminum sulfate at removing As^{+3} and at removing As^{+5} at pH > 7.5, preventing the formation of coagulant residues when the pH is above 8. One limiting factor in the removal of As^{+5} is low coagulant dosage. Ferrous sulfate is more effective and has advantages over ferrous chloride because the flocs (of good quality) of iron hydroxide formed with the ferrous sulfate assist filtration, preventing the need for frequent washing of the filters.

Fe^{+2} is easier to use and control than Fe^{+3} in the coagulation process as less maintenance is required, operation is easier and it is more economical. A hydrogel (aluminum hydroxide) made in Germany and based on hydrated aluminum sulfate, calcium hypochlorite powder, ammonium hydroxide and distilled water, can remove almost all the arsenic (< 0.01 mg/l). Removal of arsenic As^{+5} with lantane III salts depends on the pH and coagulant dosage. The optimum working conditions lie within a pH range of 5 to 10 with a coagulant dosage three times greater or equal to the concentration of As^{+5}. Residual concentrations of less than 0.010 mg/l and removal of As^{+5} greater than 83% have been obtained with pH in the range 6.54-10.6 and a reaction time of five minutes. Removal of As^{+3} takes place within a narrow pH range, with efficiency less than 60%.

The advantages of using lantane salts as coagulants are their abundance, low cost, wide pH range at which they work, the fact that small coagulant dosages can be used, its low residual concentration at optimum dosages and rapid reaction time. Arsenic removal in situ. Arsenic in underground water is mobile in reducing conditions but it can be fixed by creating oxidizing conditions at the surface of the aquifer.

An underground aquifer has been used as a natural biochemical reactor which does not generate waste or contaminating sludge. The technique is ideal for arsenic, iron, manganese, ammonium and other organic substances. According to Ahmed and Rahman, under reducing conditions and in the presence of sulfate, the arsenic can be precipitated in the form of insoluble arsenic sulphides.

Coagulation with iron and aluminum salts have been proved to be efficient at removing arsenic from a water body in two situations: a) with solid formation in situ, and b) with the addition of preformed hydroxides to the water where removal of As^{+5} in situ is five times more efficient than by using pre-formed hydroxides. This suggests that in situ the mechanism for removing arsenic is through solids with a large surface for adsorption or by co-precipitation.

In Germany, 17 wells drawing water from an aquifer contaminated by a high level of arsenic, iron (II) and having a low pH were treated by injecting 29 tons of potassium permanganate, oxidizing the As^{+3} to As^{+5}, which was precipitated and adsorbed by the ferric oxides that were formed. The arsenic concentrations fell from 13.6 to 0.060 mg/l. During another experiment in Germany, part of the water (having high levels of arsenic, iron and manganese) was returned to the aquifer containing atmospheric oxygen as an oxidizing agent. With this procedure arsenic levels of 0.005 mg/l were obtained together with low levels of iron and manganese.

Presence of Iron and Manganese

The geochemistry of arsenic reveals that high concentrations of arsenic in underground water are frequently associated with high concentrations of Fe^{+2} and Mn^{+2}. Sources of water containing iron and/or manganese and arsenic can be treated using conventional procedures for removal of Fe/Mn. These processes can significantly reduce the arsenic by removing iron and manganese from the water source, based on the mechanisms that occur when iron is added. When the natural concentration of Fe/Mn is not high enough to remove the arsenic, it can be added.

Oxidation to remove Fe^{+2} and Mn^{+2} leads to the formation of hydroxides that remove soluble arsenic by co-precipitation and adsorption. The production of oxidized types of Fe-Mn and subsequent precipitation of hydroxides are analogous to coagulation in situ (the quantity of Fe or Mn removed is equivalent to the coagulant dosage). With unoxidized Fe^{+2} or Mn^{+2}, arsenic is not removed. Each mg/l of Fe^{+2} removed is capable of adsorbing 83% of a concentration of 0.022 mg/l of As^{+5}, leaving 0.0035 mg/l of arsenic remaining.

Precipitation of 3 mg/l of Mn(II) may result in a residual concentration of 0.00375 mg/l of arsenic in a source containing 0.012 mg/l. For groundwater with high dissolved iron content, the traditional treatment of aeration and sand filtration frequently reduces arsenic to the level suggested in current standards. Adding an oxidant (oxygen, hydrogen peroxide or potassium permanganate) to a reactor containing water contaminated with ferrous sulfate (considering the processes involved in the treatment: addition of reagents, mixing, precipitation and solid/liquid separation) which contains an inert medium in the form of quartz sand grains, removes iron and other metals and non-metals, including arsenic at residual concentrations below 0.0005 mg/l.

This removal of arsenic improves as the pH decreases. The optimum treatment conditions are simultaneous addition of ferrous sulfate and hydrogen peroxide and reduction of the pH to below 7. The advantage of this method is that no waste is generated, just granules having a density of approximately 3 kg/l.

Granular Ferric Hydroxide

Granular ferric hydroxide is a β-FeOOH that is slightly crystalline. It is prepared in a suspension of iron hydroxide whose irregular granules can reach 2 mm (surface of the particles of granular iron hydroxide), adsorb As^{+5} in processes that are almost independent of the pH, with the adsorption capacity falling as the pH increases (although good results have been obtained at a pH of 8).

Compared to activated alumina, this technique is much more efficient since the hydroxide charge required by the granular iron is ten times higher (average charge around 2 g As/kg dry weight). In addition to arsenic this surface also adsorbs carbonate, silicate, fluoride and phosphate. Comparing this method with flocculation shows that it is highly reliable in operation, uses minimum energy and requires low levels of investment in the plant, with the disadvantage that the production levels are relatively expensive. Stability of As^{+3} in goethite structures (α-FeOOH) has been evaluated using X-ray absorption spectroscopy techniques and the results show that the reactivity of the As^{+3} with the goethite is independent of the pH, while bidented As^{+3} complexes were formed, similar to other oxyanions that can also be adsorbed.

The results obtained from the X-ray absorption fine structure spectroscopy show that the arsenic forms a bidented binuclear complex on the surface of the goethite with a As^{+3}–Fe distance of 3.38 Å. These results are important for modeling the transport of contaminant oxyanions, as it will be possible to forcibly calibrate chemical models to describe the formation of the known surface complexes. It has also been found that both As^{+3} and As^{+5} are removed by iron oxyhydroxide in the form of dense granules containing 50 g of arsenic per kg or more.

Iron with Direct filtration

This process consists of the addition of iron (coagulation) followed by direct filtration (microfiltration system), which consistently removes arsenic to a concentration of 0.002 mg/l. The critical parameters are the iron dosage, mixing power, retention time and pH.

Activated Alumina

The technique is effective for treating water with a high content of dissolved solids. Nevertheless, selenium, fluorine, chlorine and sulfates at high levels may compete with arsenic for adsorption sites. Activated alumina is highly selective for As^{+5} and its efficiency at removing arsenic is greater than 95%. Arsenic removal occurs under moderately acidic conditions (pH 5.5-6), where the alumina surface is protonated. Its high selectivity is a problem for the regeneration (loss of 5 to 10% of adsorption capacity/cycle) of the treatment surface.

This method is inefficient for complete removal of arsenite because of its non-ionic nature in the normal pH range of natural water. Water with a high content of solids, iron and manganese require pre-treatment to prevent obstruction of the medium. It is therefore recommended for treating groundwater which does not have the high solids content that surface water can contain. As 20-50% of the arsenic in groundwater is in the form of arsenite, a prior oxidation process is required and this can be carried out using sodium hypochlorite. If the water contains iron, this will be oxidized favouring the formation of insoluble iron compounds which adsorb the arsenic that can be separated from the water by sand filtration; the remaining soluble residual arsenic is adsorbed by activated alumina.

Ion Exchange

Ion exchange resins for removing arsenic can be weak base or strong base. Strong base resins can remove arsenate from water leaving it with less than 0.001 mg/l of arsenic. This method does not remove arsenite and allows analytical differentiation between arsenic compounds. The main materials interfering with the process are sulfate and total dissolved solids, while iron and manganese can cause obstruction of the bed. When these materials are present at high concentrations, the water must be pre-treated. Removal of the arsenic is relatively independent of the pH and initial concentration, and is almost complete (85% to 100%).

The advantages of ion exchange resins are easy regeneration using sodium chloride, wide pH range and the improvement in water quality through removal of chromate, selenate, nitrate and nitrite. This method is relatively costly and regeneration of the resin produces salt solutions that are rich in arsenic. Pilot studies for the removal of As^{+5} and nitrates from drinking water have been conducted in McFarland and Hanford, California, Albuquerque and New Mexico using sulfate selective resins and with initial arsenic concentrations of 0.010-0.015 mg/l which resulted in residual concentrations below 0.002 mg/l as well as low levels of nitrates.

Physical Separation

Some synthetic membranes can act as molecular filters to remove arsenic and other dissolved particulate compounds, as they may be permeable to certain dissolved compounds but exclude

Reverse osmosis and Nanofiltration

This provides removal efficiencies of the order of 95% when the operating pressure is at 1 psi from the ideal (75 to 250 psi). Removal of arsenic is independent of the pH and the presence of other solutes. The use of membranes requires that the water should not contain excessive quantities of colloidal material, especially organic material. With nanofiltration, arsenic removal efficiency reaches 90%.

The removal of inorganic contaminants including arsenic is achieved using reverse osmosis systems at high (400 psi) and low (200 psi) pressures with a water treatment capacity of 1.82 l/s. These systems remove between 91 and 98% of As^{+5} in high pressure reactors and between 77% and 87% in low pressure reactors. The removal efficiency for As^{+3} in high pressure systems is 63% to 70% and in low pressure systems 12% to 35%. With different types of membranes removal of As^{+3} and As^{+5} varies between 46% and 75% for initial concentrations of As^{+3} of 0.04-1.3 mg/l. At concentrations of 0.11 to 1.9 mg/l using the same type of membrane, higher efficiency in removal of As^{+5} is obtained (98% to 99%). The effectiveness of nanofiltration and reverse osmosis membranes at removing As^{+3} and As^{+5} at an operating range of 40-400 psi, using synthetic and natural water samples varies between 96% and 99%.

This separation is attributed to the high molecular weight of arsenate and arsenite, rather than rejection by load. Therefore, these membranes are ideal for groundwater where As^{+3} predominates since no prior oxidation is needed. Variations in pH (4-8) do not affect the removal of these forms of arsenic. When acetate-cellulose membranes are used, the pH range should be between 5 and 6.5 to avoid deterioration through hydrolysis of the polymer. Removal of arsenic is independent of the presence of other solutes and better results can be obtained at low temperatures.

Finally, nanofiltration membranes at operating pressures of 40 to 120 psi are as efficient as those of reverse osmosis at pressures of 200 to 400 psi. The main disadvantages are low water recovery rates (10-20%); the need to operate at very high pressures; high operating costs; the fact that treated water has very low levels of dissolved solids, which makes it corrosive; and low levels of micronutrients that are important for human health.

Reverse Electrodialysis

Removal efficiency can reach 80%. From water with a concentration of 0.021 mg/l of arsenic, a residual concentration of 0.003 mg/l was obtained. The percentage recovery of treated water is 20% to 25% with respect to the original flow. This is a problem in regions where water is scarce. This technique cannot compete with reverse osmosis and nanofiltration in terms of process cost and efficiency.

Biological Processes

Bacterial activity may play an important role as a catalyst in several of the processes used to remove arsenic, but little is known of the viability of biological processes in eliminating arsenic from water. It has lately been advanced as an alternative process for removing arsenic. Small-scale studies show that optimum conditions of pH, temperature and oxygen enable biological filtration and the simultaneous elimination of As^{+3} and iron. The critical parameter is the initial concentration of iron. At higher concentrations, the efficiency of arsenic removal reaches >90% and at lower concentrations efficiency is approximately 40%. For water systems with low iron concentrations, the addition of ferrous sulfate is recommended to complete removal of arsenic.

Fixation of As^{+3} in the iron oxides produced by bacterial activity is the main mechanism. Biological filtration in the treatment of arsenic can be applied to any groundwater system for the bacteriological oxidation of iron. Methods of filtration using spores to remove arsenic are also being successfully developed.

STUDIES AND EXPERIENCE IN LATIN AMERICA

Argentina

Studies have been carried out into removing arsenic using chemical coagulation at the Pompeya water treatment plant in San Antonio de los Cobres, department of Los Andes. The water, with an arsenic content between 0.27 mg/l and 0.30 mg/l, was treated with aluminum sulfate, hydrated lime and sodium polyelectrolyte and arsenic concentration was reduced to 0.12-0.15 mg/l. To reduce arsenic to acceptable concentrations, the plant infrastructure and treatment processes (coagulation, flocculation, sedimentation, filtration, disinfection) would have to be improved.

A direct coagulation treatment onto filters was evaluated in the province of Santa Fe, by laboratory testing and field trials. A pilot filter using water from a well chosen for its high arsenic content (0.27 mg/l) was set up; the result was a reduction in the arsenic concentration in the water.

Arsenic was found to be present in the groundwater used to supply 15 large towns in the province. In the province of Buenos Aires, water has been treated successfully since 2002 in a pilot plant using oxidation and coagulation with ferric chloride.

Mexico

The Instituto Mexicano de Tecnología del Agua has adopted a methodology based on the coagulation - flocculation process (ALUFLOC developed by CEPIS-BS/SDE/PAHO) for the removal of arsenic from drinking water. One of its objectives was to determine the optimum dosage coagulant aid, coagulant and oxidant; the resulting water had an arsenic concentration very similar to the level given in the standard (0.05 mg/l). In Zimapan, Hidalgo, a pilot study of arsenic removal was carried out. Preliminary research identified hematites as an alternative sorbent for removal. The study was carried out on well V of Zimapan, which provides 47% of the water used by the population; the water contains 0.5 to 0.9 mg/l of arsenic.

The results obtained in the field confirmed the effectiveness of hematite for removing arsenic and a concentration of less than 0.05 mg/l was found in the water treated. Various materials with a high iron content and positively charged surface were identified as alternative arsenic adsorbents. The adsorption capacity of various materials was evaluated using a sample of synthetic water containing 1 mg/l, and iron-manganese, hematite, manganese oxide and activated carbon with copper sulfate removed up to 100% of the arsenic. The laboratory studies used hematite as the adsorbent. An efficiency of 100% was observed when working within a pH range of 6 to 7, thus avoiding iron solubilization and encouraging arsenic adsorption.

At Zimapan a pilot plant was designed with a twin filter operating system using hematite and zeolite, which reduced the concentration from 0.62 to 0.875 mg/l in the untreated water to < 0.05 mg/l in the treated water. A study carried out in Mexico gave results for arsenic removal using $Al_2(SO_4)_3$ as a coagulant with solid materials that favoured the formation of flocs. Improvements also result from the use of polymers. Six solid materials were used: natural green clay, two activated clays, natural zeolite, activated zeolite and bone charcoal.

The tests were carried out on samples of synthetic water having an arsenic concentration of 1 mg/l, in a ratio of 30:70 As^{+3}:As^{+5}. The jar test was used to find the best dosage of coagulant, between 70 and 80 mg/l. Better results were obtained with 500 mg/l of the natural green clay, which produced the lowest concentration of residual arsenic (0.049 mg/l). Tests were carried out using natural water from Zimapan, and the best dosage was found to be 80 mg/l of aluminum sulfate, 100 mg/l of green clay and 0.5 mg/l of calcium hypochlorite (oxidant).

A final report by PAHO for the state of Durango proposed techniques for rural and urban areas. In rural areas it was proposed to use ground bone and lime-aluminum sulfate. In urban areas, especially the central area of Mexico, alternative sources were proposed together with mixtures of water, ground bone, activated alumina and reverse osmosis. A study was also carried out on removing arsenic dissolved in water by adsorption of arsenic ions in metallic precipitates in an electrical treatment.

The technique used iron electrodes. Samples of synthetic water in concentrations of 0.5 to 2 ppm were treated and the voltage was between 13.5 and 14 V. The Fe/As ratio at which removal efficiencies in excess of 90% were achieved was in the order of 225 to 700. The pH varied between 6 and 8 and did not change significantly at the end of the experiment; the quality of the water produced was very good.

Chile

A regulation has been prepared for maximum levels of arsenic in drinking water, taking into account local conditions, maximum levels of emission from foundries, areas of influence and technological options for significantly reducing these emissions. In the 1970s measures started to be taken to remove arsenic from surface water sources supplying the people of Antofagasta.

At present there are four plants removing arsenic from the supply water (0.40 mg/l) which together treat 2,000 l/s and produce drinking water with an arsenic content of 0.040 mg As/l; only the scattered population consisting mainly of indigenous groups have not been covered and still drink water with a high concentration of arsenic (0.6 mg/l). For these people, simple methods of removing arsenic have been looked into and solutions are currently being adopted according to their needs and possibilities.

For the risk analysis in Antofagasta and Calama, various options for removing arsenic from the drinking water supply were examined, together with the associated costs. The study made a technical and economic evaluation of the best treatment system, adding reverse osmosis as further treatment of the water treated by the existing system and installing a desalination plant, the product of which is mixed with the water produced under current treatment conditions. This would imply reducing the residual arsenic in the drinking water to 0.020 mg/l in the short term and at a very low cost. The Chilean experience in removing arsenic is summarized below:

- Removal from groundwater supplying Taltal (8,000 inhabitants) is carried out through direct filtration with the addition of $FeCl_3$ coagulant, which reduced the arsenic content to 0.03 mg/l.
- In Lasana, direct filtration and the addition of $FeCl_3$ coagulant was used to reduce the arsenic content from 0.05 to 0.005 mg/l. Iron mesh was also used as a source for the coagulant and the arsenic content was reduced from 0.1 to 0.01 mg/l.
- At Antofagasta (Salar El Carmen complex), arsenic was removed by direct filtration with the addition of $FeCl_3$ coagulant; the residual arsenic in the treated water was 0.0406 mg/l.
- At Chiu-Chiu, direct filtration and the addition of FeCl3 coagulant was employed, which reduced the arsenic content from 0.2 to 0.62 mg/l. The use of iron mesh as coagulant reduced the arsenic content from 0.2 to 0.02 mg/l.

- At San Pedro de Atacama, columns of iron mesh were used followed by filtration through a simple coarse sand and fine sand filter bed, which reduced the arsenic content to 0.06 mg/l.

Treatment plants for arsenic removal in Chile have obtained very similar results to those obtained by Cheng *et al.* In this case aluminum sulfate and ferric trichloride were used; arsenic removal was much better using $FeCl_3$ and was less dependent on the pH. The factors affecting arsenic removal in this study were: the pH of the water to be treated, the coagulant dosage and the process of separating the flocs that were formed.

The arsenic removal processes can be improved by optimizing the pH of the raw water and chemical agents, and floc separation, which increases removal of arsenic to 92.5-95%. In order to obtain residual arsenic values lower than 0.020 mg/l in the treated water, more advanced treatment techniques should be used, such as further treatment by reverse osmosis.

A later study evaluated the effects of pH and the water source, finding that removal was more efficient at a pH of 6.5 and coagulant dosage of 10-11 mg/l of $FeCl_3$, which produced very good adsorption of arsenic and removal of the flocs. An empirical formula was also arrived at to predict the residual arsenic under different operational conditions, and this was confirmed by data obtained in the treatment plant.

Peru

The treatment plant in the city of Ilo was built in 1982 and designed to eliminate arsenic and turbidity. Treatment initially used massive doses of 90% lime, but many difficulties were encountered and it was necessary to use new methods of treatment to reduce costs, increase efficiency and improve the quality of the treated water. In order to eliminate arsenic, three methods were examined and applied in the plant using ferric chloride; with ferric hydroxide and sulfuric acid, removing arsenic at high pH through coagulation and flocculation with natural $Mg(OH)_2$.

CEPIS-BS/SDE/PAHO

A product known as ALUFLOC was developed, being a mixture of an oxidant, activated clays and a coagulant (aluminum sulfate or ferric chloride). To mitigate this problem a simple, low-cost methodology has been developed which enables household removal of the natural arsenic present in the groundwater which is used as drinking water by the rural population.

Up to 98% of the arsenic was removed using $FeCl_3$ as the coagulant, when the initial arsenic concentration was 1 mg/l. For concentrations lower than 1 mg/l, the use of $Al_2(SO)_3$ is recommended. The best mixtures of clay, coagulant and oxidant were: activated clay A (500 mg/l), aluminum sulfate (50 mg/l) and chlorine (5 mg/l), and activated clay B (1,000 mg/l), ferric chloride (60 mg/l) and chlorine (50 mg/l). Three types of reactors were tested with the best mixers being a vertical axis bucket and a 20 l bottle with paddle instead of the reactor with hand wheel. Finally, the study recommended to use the first mixture for household arsenic removal because it is easy to handle.

When the arsenic concentration in the untreated water is greater than 1.0 mg/l, removal diminishes as the initial concentration increases, particularly if aluminum sulfate is used. Coagulation using aluminum sulfate or ferric chloride does not remove As^{+3} as efficiently as it removes As^{+5}. The first requires preliminary oxidation before it can be removed using conventional coagulation, and/or softening with lime and/or soda.

After preliminary treatment with chlorine, As^{+3} removal is similar to that of As^{+5} with the same treatment, because the dissociation constants of As^{+5} are lower than those of As^{+3} and therefore its dissociation is greater. The efficiency of As^{+5} removal compared to As^{+3} justifies the oxidation of groundwater before treatment. The oxidation of As^{+3} to As^{+5} occurs through the following reactions:

$$H_3AsO_3 + H_2O + [O_x] \rightarrow H_2AsO_4^- + 3H^+$$
$$H_3AsO_3 + H_2O + [O_x] \rightarrow H_2AsO_4^- + 4H^+$$

After oxidation, As^{+5} must be adsorbed in the flocs formed by coagulation of colloids present in the water through the action of the aluminum or iron hydroxide. Adsorption probably results more of a physical effect than a chemical one, given the surface attraction forces which make the phenomenon possible.

Probable removal mechanisms: The mechanisms that probably govern the removal of arsenic (solute) by clays and/or metallic hydroxides (sorbents) depend on molecular interactions in the arsenic/water/hydroxide-clay system, where linkages are established that allow the active surfaces to change the arsenic compounds present in the water as a result of intermolecular chemical reactions between the arsenic and the active surface and between the arsenic and the water.

The greatest contact possible between the two phases must be ensured to favour adsorption. Therefore the reactions between the arsenic and the surface, through coordinated linkages formed by hydrolysis, surface change, exchange of bonding agents and hydrogen bonding are of fundamental importance. This latter is weak compared to other chemical reactions, but is important for the adsorption of molecules by the polarized hydroxides.

The electrical interactions on the surface (electrostatic and polarization) are also important, as are the interaction between the arsenic and the water (dependent on the pH and pE of the medium). On the surfaces, energy freed by adsorption can be represented by:

$$\Delta G_{ads} = \Delta G_{chm} + \Delta G_{elect} + \Delta G_{solv}$$

where:

ΔGchem = energy associated with the hydroxide/clay-arsenic-water chemical reactions, which are a function of the chemical interactions between the water and arsenic; water and clay and/or hydroxide; and the clay and/or hydroxide with the arsenic.

ΔGelect = energy required to carry the arsenic to the active surface, a function of electrostatic interactions.

ΔGsolv = energy required for solvation or hydration, an exponential function of the ion charge.

Arsenates hydrate much less than the cations, therefore the change in the pH of the water has a significant effect on the adsorption of arsenates because the anion forms in the water have different solvation energies.

The probable mechanism for removal by adsorption through the surface of the clay-hydroxide complex

- In an aqueous system the surface of the clay-hydroxide floc may contain hydroxyl groups, S-OH, produced by the coordination of metallic ions with water. These groups can gain protons from the water, giving the surface a positive charge, or lose protons giving the surface a negative charge. The ions which determine the surface potential of the unpolarizable hydrated oxides are hydronium (H_3O^+) and hydroxyl (OH^-). Changes in the structural group OH^- -may occur on the surface of the floc because of the reaction with arsenates or dissolved bonding agents, so that they display the properties of a strong acid (Lewis acid). The exchange of bonding agents can involve changes in the surface of the clay-iron or aluminum hydroxide flocs, OH^-, as in the reactions:

  $$S{-}OH + L \longrightarrow S{-}L^+ + OH^-$$

  $$2S{-}OH + L \longrightarrow S_2L^{+2} + 2OH^-$$

  $$S_2L^{+2} + HAsO_4^{-2} \longrightarrow \text{adsorption of the } As^{+5} \text{ in the floc}$$

Adsorption of arsenates in the clay-hydroxide system will depend on the density of the OH^-groups and their selectivity with respect to the other water anions. Another possible way that arsenic is removed from the water is adsorption of arsenates of the same compound through ions from other dissolved elements, which produce ternary surface complexes such as:

(*i*) $S—OH + M^{2+} + L \longrightarrow S—ML_1^{(z-1)} + H^+$ $S—ML_1^{(z-1)} + AsO_4^{2-}$ adsorptionof As^{+5}

Where the adsorbed metallic cation does not coordinate completely with the surface bonding agents and is completed with other bonding agents dissolved in the water. In this case the cation operates as a "bridge" between the surface and the S-OM-L bonding agent, which could produce adsorption of the arsenate by the bonding agent.

(*ii*) $S—OH + L + M^{z+} \longrightarrow S—L—M^{(z-1)+} + OH^-$ $S—L—M^{(z+1)+}$ + $AsO_4^{-2} \longrightarrow$ adsorption of $As+^5$

Where the bonding agent L is directly adsorbed into the centre of coordination S and the cation completes the surface coordination. Complexes of these types are formed with polydented L bonding agents, which form a "bridge" between the surface and the S-L-M cation and this adsorbs the arsenate by electrostatic action. The resulting surface charge and pH are determining factors in later adsorption and in the clay-hydroxide/arsenate interactions respectively.

Because of the chemical transformation of the surface of the sorbents, they may change their properties and ability to remove arsenic in the chosen media.

- What will have to be evaluated is the reaction between the negatively charged clay and the clayhydroxide surface with functional OH^- groups, because this may compete with the adsorption of the arsenates. These evaluations should take into account that activated clays are good ion exchangers, they behave like a solid acid and have an average pore volume of 600 Å.

The chemical adsorption reactions between the clays and the hydroxide OH groups should also be evaluated. Conditions for maximum efficiency. The iron $Fe(OH)_3$ and aluminum $Al(OH)_3$ precipitates are the end products of hydrolysis of the coagulants $Al_2(SO_4)_3$ and $FeCl_3$ dissolved in the water and relatively larger quantities are formed at a pH of 6.5 to 7.0 for aluminum and a pH of 7.5 to 8.0 for the iron.

The $Fe(OH)_3$ precipitate is almost insoluble at a pH of 3.0 to 13.0 and, in alkaline conditions precipitates in accordance with the following reaction: $Fe^{+3} + 3OH^- = Fe(OH)_3$. The $Fe(OH)_3$ floc at pH < 6.5 has a positive surface charge and at pH > 8, a negative surface charge. In the pH range between 6.5 and 8.0 the surface charge is mixed.

$Fe(OH)_3$ is least soluble at pH 8.0. The soluble forms of iron at pH < 6.0 are Fe^{+3} having a higher concentration at pH < 3.0 and $FeOH^{+2}$ predominates at pH < 4.0 while the $Fe(OH)_2^-$ concentration is highest at a pH between 5.0 and 7.0. At pH > 10, the predominant soluble form is $Fe(OH)_4^-$.

At a pH of < 6,0 the predominant forms are Fe^{+3} $FeOH^{+2}$ and $Fe(OH)_2^+$ which favour the adsorption of arsenate ions, but in this same range the adsorption capacity of the arsenates by ferric hydroxide is limited by its negative surface charge. This can be explained because better arsenic removal is obtained with ferric chloride when the pH is in the range 7.0 to 7.5, where removal of arsenic takes place through adsorption and co-precipitation of both cation forms and $Fe(OH)_3$ hydroxide precipitates, with which the arsenic co-precipitates and sediments rapidly.

The precipitate of $Al(OH)_3$, which in reality is $AlO_3 . \times H_2$ has a positive surface charge at pH < 7.6 and negative at pH > 8.2. In the pH range between 7.6 and 8.2 the surface charge is mixed. At a pH of 4.0 the Al^{+3} forms predominate, whilst at a pH of 9.0 anionic $Al(OH)_4^-$ forms predominate. In all cases, removal of arsenic by coagulation and adsorption must take place with greater efficiency at a pH < 7.0 when the most effective cationic forms of iron and aluminum are present and competition from the hydroxyl group is reduced.

SOIL TREATMENT

Phytoremediation

In recent years phytoremediation has proved a novel, efficient and economic technique for the recovery of contaminated soils. Contamination of soil by heavy metals, in this case arsenic, is difficult to eliminate and remains for decades. It is concentrated along the food chain poisoning and causes malformations in the higher animals. In recent years a new technology for eliminating contaminants from affected soils has been developed, a technology which is clean, cheap and effective.

This technique is known as phytoremediation and consists in planting contaminated areas with certain plant species which are known to absorb and concentrate toxic substances. These species are referred to as hyperaccumulators. The plants can help to clean various types of contamination including metals, insecticides, explosives and oil; they remove the contaminants from the ground when their roots absorb water and nutrients from contaminated soil and groundwater. Plants are better at cleaning contaminants from the soil when they have deep roots. The plant may store the contaminants in their roots, stems and leaves; converted into less damaging contaminants within the plant; or converted into gases that are released into the air when the plant transpires.

The time taken to clean the area depends on various factors such as the species and number of plants used; the type and quantities of contaminants present; the size and depth of the contaminated area; and soil characteristics and conditions. These factors vary at different sites. Plants can be replaced if they are destroyed by the weather or animals. It will therefore take time to clean this area.

Cleaning land by phytoremediation frequently takes several years. The EPA uses phytoremediation because of the advantages it offers. It requires less equipment and work than other methods, as the plants do most of the work. A site can be cleaned without having to remove the topsoil or pump out contaminated groundwater. This prevents workers coming into contact with the contaminants. Phytoremediation has been proven satisfactorily in various places and may be used on contaminated sites.

A Fern to Remove Arsenic

A team of scientists from the University of Florida have discovered a fern that absorbs arsenic from contaminated soil. This is the first plant found to be a hyperaccumulator of arsenic, that is, it uses arsenic as part of its food. The fern, whose scientific name is Pteris vittata, is easy to grow and prefers sunny places with alkaline soil. This latter characteristic favours the absorption of arsenic, which can easily be extracted from alkaline soils. Studies have shown that this fern is highly efficient at removing arsenic from soils.

Tests have found levels up to 200 times higher than the concentrations measured in contaminated soils where the fern is growing. A site contaminated by timber treated with a solution of chromium, copper and arsenic, the arsenic concentration in the soil was 38.9 mg/k whilst that in the fern was 7.526 mg/k. In tests using artificial soil contaminated with arsenic, the concentration of this metalloid in the fern's leaves reached 22,630 mg/k of arsenic, which means that 2.3 per cent of the plant consisted of arsenic. It has been shown that the fern accumulates arsenic in soils that contain normal levels of less than 1 mg/k.

For example, a fern contained 136 mg/k in its leaves when the soil contained only 0.47 mg/k of arsenic. It appears that the plant grows and develops much better in soils containing arsenic than in those that do not, although it is not clear yet whether the plant needs arsenic to survive. Because the fern accumulates 90% of the arsenic in its leaves and stems, the strategy would be to plant it in contaminated areas, then remove the leaves and stems to a facility authorized to treat hazardous waste. What is not yet known is why the fern accumulates arsenic. Future research will concentrate on how the plant absorbs, distributes and detoxifies the arsenic.

Treatment of Sediment and Sludge (waste)

Treatment technologies for removing arsenic and industrial activities generate waste with a high arsenic content. This has led to a search for solutions to recover and stabilize the arsenic or to minimize the high risk of contamination that it represents. There are various technologies for this purpose. Hydroxide precipitation is effective and economical at removing heavy metals, the precipitates being separated by sedimentation and/or filtration. Lime and caustic soda are generally used for this. The disadvantage is the quantity of waste generated and how to dispose of it. Another option could be precipitation with sulphides, which is highly efficient at arsenic removal.

The compounds thus formed are insoluble. For the removal of arsenic to be more efficient, the process should take place at pH < 7. Nevertheless its use has been limited because of the toxicity and smell of the H_2S. Chemical fixation and solidification offer many advantages, such as improving the handling characteristics, producing a solid material that is sufficiently stable for disposal. This technology is used to detoxify, immobilize and make insoluble or, in other words, to make the waste less dangerous for the environment.

A study has been carried out in Mexico into making an insoluble compound of arsenic which can also be used as the raw material for making other solid compounds that can be used in construction or disposed of in a landfill site. An examination of state-of-the-art removal of arsenic from drinking water indicates that technology transfer relating to improving water quality in developing countries is especially important, and should consider:

- The characteristics of the sources, its adaptation, means of distribution and/or consumption and the different technologies that should be applied under different circumstances, taking into account the distinctive characteristics of each location.
- The cost/benefit of using the proposed technology should also be considered, as it should clearly solve a social, public health or development problem affecting the community in question.
- Existence of trained personnel to help in the development of projects and later maintenance and monitoring.

- The understanding and acceptance of the proposed technology by the community, which should include its capacity to finance its implementation.

As far as soil remediation is concerned, recent experiences and preliminary information should be followed and studies carried out into the application of these techniques under different conditions (soil, environmental, cultural, final disposal of waste, impact on ecosystems, etc.).

Latin American countries have the experience and ability to develop the technology, but are limited by a lack of financial resources, facilities and above all state policies to facilitate and direct the development of the technology leading to an effective solution or satisfaction of existing needs. Those most affected are scattered over rural areas, they consume untreated water and are unaware of the risks to which they are exposed.

These people need the health, planning and water authorities, among others, to promote and implement programs to prevent and control the risks of consuming water with arsenic levels that are higher than those recommended by drinking water standards. These programs must involve the participation of the authorities, the community and the local health systems.

It is necessary to conduct continual and sustained pilot projects until a final solution is developed that can be recommended for implementation in national programs for removing arsenic from drinking water. Studies that go beyond laboratory work are required, together with epidemiological studies and use of experiences in Latin America. A reliable and comparative analytical capability must be developed in order to obtain reliable results in the studies carried out at the laboratory and field level.

Chapter 9

Environmental Protection through Biodiesel Plantations

Abstract: Increasing energy demand and spiraling oil prices are the cause of financial strain on countries and also causing environmental degradation. Use of non-edible oil as biodiesel provides a win-win proposition for the densely populated Asian countries. Perennials such as Pongamia pinnata and Jatropha curcas have offered an excellent opportunity to remedy the problems of enrionmental protection and oil crisis.

Strategies for rehabilitating degraded lands and improving livelihoods through biodiesel plantations based on field experiences are discussed. Although technology, policy support, and the demand for biodiesel are increasing, the main constraint is supply of raw material.

Validated and good quality data on agronomic practices, yield potential, diseases and pests occurrence, water requirement and management for block plantations are not available. There is an urgent need to undertake research on all aspects of biodiesel plantations. Energy security has assumed a greater significance than ever before as energy consumption, food production, improved livelihoods and environmental quality are interrelated.

Asian countries with dense population are more prone to energy crises than are their counterparts in the developed world. A strong nexus between energy use for irrigation water and agricultural output exists. The number of borewells in India and energy consumption have increased dramatically over the last two decades.

Groundwater now sustains almost 60 per cent of the India's irrigated area. Even more importantly, groundwater now contributes more to agricultural wealth creation than any other irrigation source. During a recent study undertaken by ICRISAT in Rajasamadhiyala watershed in Gujarat it was noted that not only had the number of open and bore wells in a village of 1000 ha increased from 357 to 508 within eight years but pumping hours had also increased from 5.25 to 10.4 hours per day, putting pressure both on energy resources and groundwater.

In addition, farmers also keep additional diesel pumpsets as a stand by to cope with the previailing power cuts. Agriculture is now more dependent on mechanical and electrical sources of energy than on human and animal power. During 1988-89 to 2004-05, farm power availability in Anantapur Village, a predominantly dryland area increased by 28 per cent, while mechanical power use increased remarkably by 730 per cent.

Any increase in food production c alls for higher energy use in terms of irrigation and fertilizer, as further expansion of area under agriculture is limited. Areas that are left fallow or degraded and not fit for agriculture offer an opportunity for greening through tree-based oilseeds (TBOs).

The advantages of perennials are many for example, greenery will protect the land from further degradation and generate employment in rural areas. The total number of species with oleaginous seed material mentioned from different sources varies from 100 to 300 and of these 63 belonging to 30 plant families hold promise. Two species namely, Jatrohpa curas and Pogamia pinnata are favoured in India because of their contrasting plant characteristics and the species selected should match the site characteristics.

JATROPHA

Jatropha is highly drought tolerant and is well suited to semi-arid conditions although it thrives in arid areas. It occurs mainly at lower altitues (0-500 m) in areas with an average annual temperature well above 20 °C but can grow at higher altitudes, tolerates slight frost and up to 1000 mm rainfall.

Pongamia thrives well up to 1200 m altitude and rainfall of 500-2500 mm per annum. Pongamia withstands temperature from about -1°C to 50°C. Both plants can withstand drought but Jatropha sheds its leaves during the dry period (Summer) while Pongamia retains leaves and is evergreen.

Gall formation on leaves caused by mite in Pongamia is a major concern as it affects leaf area/photosynthesis and it is widespread. Jatropha has an affinity for well-drained soils and is susceptible to collar rot disease in high rainfall humid areas or excessive irrigation or water logging conditions, while Pongamia can tolerate water logging, saline and alkaline conditions. Both the species are suitable for semi-arid tropics and are not grazed or browsed by livesock.

Table. Jatropha vis-à-vis Pongamia

Characteristics	Jatropha Curcas	Pongamia Pinnata
Ecosystem	Arid to semi-arid, up to 500 m altitude	Semi-arid to sub-humid, up to 1200 m altitude
Rainfall	Low to medium up to 1000 mm	Medium to high up to 2500 mm
Soil	Well drained soils	Tolerant to water logging, saline and alkaline soils
Nitrogen fixation	Non fixer	N_2-fixing
Plant suitability	Wastelands, degraded live fence for arable land, green capping of bunds, dshallow soils	Field boundary, river and stream banks stabilization wastelands, tank foreshore
Plant habit	Mostly bush, can be trained as small tree	Tree can be managed as bush by repeated pruning but will affect yield
Leaves livestock	Not palatable by used as green leaf mulch	Not palatable by livestock,
Gestation period	Short, 3rd year	Long, 7th year
Yield	1.0 kg/plant	10 to 200 kg/tree
Oil content	27-38% in seed	27-39% in kernel
Protein	38%	30-40%
Fire wood	Not useful	Good as firewood, high Calorific value 4600 k cal/kg

SEEDLING MANAGEMENT

Tree improvement programmes have mainly concentrated on a few plants like Eucalyptus spp and poplar (Populus deltoids) but no systematic efforts have been made in the past to improve tree borne oilseeds like Pongamia and Jatropha.

As the genotype and the environment influences yiels, therefore, the source of seed assumes greater significance. The quality of the seedlings also affects survival, growth as well as yield and this is further dependent on media and propagation technique.

Variability in Seed Sources

It is desirable to select the seed source for multiplication from known plant populations with favourable traits. However, no systematic efforts have been made to evaluate and document the existing variability in seed sources of both the plants. Since the last 2-3 years, a few researchers have started collecting seeds from high yielding trees of Pongamia and Jatropha. A preliminary study conducted at ICRISAT for samples collected in the states of India by various researchers and suppliers indicates great variability in terms of Jatropha and Pongamia oil content, 100 seed weight and per cent germination.

The seed source of Jatropha IJC-2 and IJC-6 were found to be promising in terms of oil content.

Table. Evaluation of Seed Samples of Jatropha and Pongamia for Oil Content, Test Weight and Per cent Germination

Sl. No.	Collection No.	Place of collection	100 seed weight (g)	% Oil content	Germination %
Jatropha					
1	IJC-1	Rajgarh	63.8	28.0	10
2	IJC-2	AP	68.2	38.4	90
3	IJC-3	Tamilnadu (cape virdi type)	49.2	29.9	61
4	IJC-4	Tamilnadu (Erode local)	44.0	28.6	85
5	IJC-5	Tamilnadu (wild collection)	77.2	29.3	50
6	IJC6	Rajasthan	69.5	34.8	78
7	IJC-7	Rajasthan	51.3	29.3	51
8	IJC-8	JNKV	72.6	34.7	–
9	IJC-10	CHRK-VSP	67.3	32.8	70
10	IJC-11	MONDC	69.2	31.8	–
11	IJC-11	CHRK-GBR	69.4	34.4	17
12	IJC-12	TFRI	66.5	33.5	–.
13	IJC-13	Rajasthan	57.6	34.4	30
14	IJC-14	-do-	53.2	27.8	23
15	IJC-15	-do-	60.67	33.6	42
Pongamia					
1	P-1	Mumbai	233.2	30.4	90
2	P-2	Goa	132.7	21.3	90
3	P-3	Orissa	155.0	33.5	90
4	P-4	Bangalore	177.0	40.9	90
5	P-5	Adilabad	102.4	39.5	90
6	P-6	Adilabad (Belrangugda)	105.2	32.9	90
7	P-7	Gurgaon	101.0	34.6	85
8	P-8	-do	101.5	33.3	85
9	P-9	-do-	101.2	36.6	85
10	P-10	-do-	102.0	33.1	85

NURSERY TECHNIQUES

There are several propagation techniques like direct seeding, cuttings, and nursery raising, which are discussed in detail below.

Direct Seeding

One to two seeds of Jatropha and Pongamia can be sown directly in the ploughed and marked field. The main field should be ploughed and spots may be marked and enriched with FYM receiving the seeds. Twenty gram urea, 120 g single super phosphate and 60 g muriate of potash should be applied after establishment of the plant.

Seedlings grown by this method will take time for establishment and will have slow growth in the initial period. However, there should be enough moisture in the soil to support germination and seedling growth. Frequent weeding is required to prevent the seedlings from competing weeds and shade. Similarly, soil working around the seedlings will boost the growth and will improve the moisture and rainfall infiltration. Pongamia seedlings grown this way can be in-situ grafted with a scion from high yielding tree after nine months.

Vegetative Propagation

Jatropha can be multiplied by raising the cuttings on a raised bed and can be later transplanted into the main field. Multiplication of Pongamia by cuttings is difficult but grafting is easier. Cuttings of 2-3 cm thickness from the lower portion of the shoot and of 25-30 cm length may be prepared in the month of March, when plants shed most of their leaves.

These cuttings are planted in nursery beds at a spacing of 30x30 cm or in polybags. Pre-treatment of stem cuttings with 300 ppm IBA (Indole butyric acid) solution for 5 minutes is desirable. Sprouting starts within 7 days of planting. Plants propagated by cuttings will normally produce seeds within a year of planting and growth is rapid.

However, it has been observed that seedlings raised from seeds have a better root system compared to pre-rooted cuttings, which develop secondary and tertiary fine roots. Such seedlings are more stable during storms. The plants raised through cuttings will be true to type and will have similar characteristics of the mother plant.

To maintain the quality of plantations of Pongamia, there is a possibility of in-situ grafting. The rootstock may be raised directly by seeding in the pits (Two to three seeds.) The desirable scion material may be grafted on the rootstock when the seedlings attain pencil thickness. The technique needs to be standardized for Pongamia as followed in mango.

Nursery Raising

Jatropha plant can be grown by two methods, namely bare root and container method (Polythene bag). In the bare root method, nursery bed is prepared by mixing FYM, soil and sand in equal volumes. Soaked seeds are sown at a row spacing of 25 cm and a plant to plant distance of 5cm. The plants will be ready for transplantation in the main field six weeks after germination.

The plants may be carefully uprooted from the nursery beds, wrapped in wet gunny bag and transported to the main field. Uprooted seedlings may be transplanted within 24 hrs. Before transplanting it should be ensured that enough moisture is available in the pit receiving the bare root seedlings. Bare root seedling production of Pongamia is difficult as it has a tap root system and will get disturbed during lifting from nursery beds. Seedlings of both Jatropha and Pongamia can be grown in poly bags of mostly black colour (4" × 7", 150 gauge for 3-4 months old seedlings). The bags may be filled with equal parts of soil, sand and FYM (1/3, 1/3, 1/3). 1.0 g per polybag of diammonium phosphate (DAP) may be added to 2 Kg soil weight. Good quality seeds having 80 per cent germination should be sown with one seed per bag at 2-3 cm depth for getting higher per cent germination.

One gram of mycorrhizae (mixed culture) may be placed below the seed at the time of sowing to enhance growth of seedlings. Jatropha seedlings treated with mycorrhize in unsterlized soil. FYM mixture had higher plant height, girth and number of leaves compared to control, when sampled 85 days after sowing. Similarly, treatment of Pongamia seeds with RhiZobium and mycorrhizae, Azatobacter was found promising for improving seedling growth. Pongamia seedlings treated with Rhizobium and Azatobacter had a higher number of nodules, shoot and root weight when sampled 85 days after sowing. Grading and root pruning is suggested to promote uniform growth of seedlings.

FIELD MANAGEMENT

Genotype and the environment (field condition) influence the survival and growth of the seedlings. There are several steps involved in management of the field.

Table. Effect of Mycorrhizal Inoculation on Growth of Jatropha Seedlings in Nursery

Treatment	Plant Height	Stem Girth	Number of Leaves
Inoculated	47	6.5	16
Non-inoculated (control)	35	5.9	12

Table. Effect of Rhizobium and Mycorrhizal Inoculation on Nodulation and Biomass of Pongamia Seedlings in Nursery

Treatment	Nodule number/ seedling	Fresh weight (g/seedling)		Dry weight (g/seedling)	
		Shoot	Root	Shoot	Root
No treatment (Control)	11	5.5	5.0	1.2	1.3
Rhizobium alone	19	6.0	5.4	1.3	1.8
Mycorrhizae alone	12	5.8	5.9	1.3	1.5
Mycorrhizae + Rhizobium	17	6.2	7.9	2.3	1.9
Mean	15	5.9	6.0	1.5	1.6
LSD ($P \leq 0.05$)	1.8	2.4	2.0	0.98	0.69

First, the field may be ploughed (deep tilled) followed by harrowing at the beginning of the rainy season or by utilizing off-season rains. Direct planting may be taken up in dug out pits in hilly and rocky areas where cultivation is not possible.

A spacing of 3m × 2m or 3m × 3m spacing is desirable for intercropping and intercultivation. For hedgerow/boundary plantation spacing of 5m × 5m or 6m × 6m is suggested. For avenue and field boundary planting of Jatropha and Pongamia and alignment of rows in the east-west direction is suggested. A spacing of 6m × 6m for Pongamia and 3m × 3m for Jatropha is desirable.

The recommended size of the pit is 30 cm × 30 cm for Jatropha and 45cm x 45cm x 45cm for Pongamia. The pits may be dug well in advance of planting time. Refilling of the pits may be done by mixing 1.0 to 2 Kg of FYM, 50g of DAP and methyl parathion (2 Per cent dust)@ 5 to 10g or 5g of thimmet granules per pit may be applied to protect the young saplings from termites damage.

TRANSPLANTING

Seeding having a plant height of 60 cm may be transplanted with the on-set of rainy season and the soil around the seedlings should be compacted. Normally, Jatropha attains a height of 60 cm in 60 to 75 days while Pongamia takes little longer time 90-105 days. The survival percentage will be more in case of plantation raised by seedlings compared to direct seeding. Seedlings are susceptible to competition from weeds in the first year. Therefore, weed control either manual or by herbicide is required during the establishment phase.

IRRIGATION AND MOISTURE CONSERVATION

Although Jatropha and Pongamia are hardy plants, even then they require adequate moisture in the root zone during the initial period.

Irrigation once in a month during the dry period will be quite beneficial for enhancing growth and productivity. Rain water conservation techniques like planting on contour or staggered trenches will be intercropping is not possible. Making ring basins around the plants and mulching (dust mulch/organic mulch) will conserve soil moisture and minimize irrigation needs.

Tree loppings and straw is ideal for conserving the soil moisture in the basins. Creating surface roughness by intercultivation, etc. will enhance rainwater conservation and use. Irrigation needs vary with local soil and climatic conditions and need to be standardized. Jatropha can stand long periods of drought by shedding leaves while pongamia remains green during dry periods. Irrigation during dry periods at fortnightly intervals during first year is suggested to improve growth and yield, if available.

However with appropriates soil moisture conservation techniques, Jatropha and Pongamia plantations in deraded common lands in Andhra Pradesh, Madhya Pradesh and Rajasthan showed 70-75 per cent survival after one year without any irrigation.

MANURES AND FERTILIZERS

It is often said that Jatropha and Pongamia do not need any nutrition. The plants may survive in soil having poor fertility but seed production will be very poor and will have only foliage. There are no systematic studies conducted on the nutritional requirement for Jatropha and Pongamia block plantations.

However, based on experience with other plantaions it may be seen that organic manures (FYM/Vermicompost) 1 Kg/plant for Jatropha and 2 Kg/plant for pongamia and a fertilizer dose of 50 g of DAP for Jatropha and 100g of DAP for Pongamia in the first year need to be applied at the time of filling the pits. In subsequent years, top dressing of Jatropha 50g and 100g of urea and SSP per plant, respectively while for Pongamia DAP 100g per plant is suggested.

Alternatively, Oil cakes can be recycled in these plants to maintain the productivity of soil. The plants respond well to addition of small quantities of calcium, magnesium and sulphur. Mycorrhizal associations have been observed in Jatropha and they are known to aid the plant growth under conditions where phosphorus availability is low.

WEEDING

The basin should be kept free from weeds. Hoeing and weeding is essential during the establishment period. Around 2-3 weedings/

intercultivations are enough to keep the field free from weeds and conserve moisture.

IRRIGATION

Jatropha was found to respond to respond to irrigation and was also recommended by the growers. Similarly, a field study in Nalgonda, revealed that Pongamia too has a liking for water and fertile soils. Pongamia grows mostly along the watercourses like streams and waterways while traditionally it is grown on field and farm bunds particularly around paddy fields.

CANOPY MANAGEMENT

The flowering occurs at the terminal portion of the branches in Jatropha and along the branches in case of Pongamia. Therefore, efforts should be made to train and prune the plant in such a way that the number of fruiting branches increase. In Jatropha, the terminal bud should be nipped to induce secondary branches. Likewise the secondary and tertiary branches are to be pinched or pruned at the end of first year to induce a minimum of twenty-five branches at the end of second year. Once in ten years Jatropha may be cut back leaving a one-foot height from the ground level for rejuvenation.

The growth is quick and plant starts yielding in about a years period. This will be useful to induce new growth and to stabilize yield. In a study of canopy management in North East Thailand, the cutting of Jatropha at a height of 50cm from the bottom was found reasonable to maintain a compact bush form. The end of the dry season was found to be the optimum time for cutting back as the plants go in dormancy after the fruiting season. Thinning twice after one and two months of cut back was recommended to promote useful fruiting branches and to maintain a compact bush form.

The Pongamia plants may be pruned initially to give the stem a straight form and later lightly lopped for green leaf mulch. All the side branches of tree, one third from the bottom may be pruned and the top two third branches on the plant may be retained. Periodical pruning can be carried out depending upon the vegetative growth of the plants. The pruning should be done when trees shed leaves and enter into a period of dormancy. Diseased, dead, excessive, weak and lateral branches should be removed.

INSECT PESTS

In the case of Jatropha, insects such as leaf eating beetles, thrips, leaf hoppers, grass hoppers, caterpillars and leaf miner will feed on foliage. Shoot/stem borer and bark eating caterpillar will damage the stem. Blue bugs and green stink bug will be sucking on fruits while capsule borer will damage the fruits.

In the case of Pongamia, leaf miner, leaf galls, bark eating caterpillar and other pests are found. The pest may be controlled by spraying Endosulfan 3 ml per litre of water or any other pesticide recommended for that particular pest.

The galls are formed due to the attack of mites and can be controlled by spraying Dicofol 5 ml per litre of water or wettable sulphur 3 g per litre of water.

DISEASES

In case of Jatropha, collar rot may become a problem in some areas in monoculture plantations under irrigated condition. It is severe under water logging conditions and excess soil moisture. The rot can be controlled by drenching with 1 per cent Bordeaux mixture. Minor diseases such as root rot, damping off, powdery mildew and leaf spots are reported. There is no specific mention of diseases in case of Pongamia.

HARVESTING

The flowering in Jatropha depends upon the location and agroclimatic conditions. Generally it takes place from August to December in India. However, this flowering and fruiting depends on the site, soil moisture and climatic conditions. Fruits mature within two to four months after flowering.

The ripe fruits should be harvested/plucked when it reaches physiological maturity (yellow capsule stage). There is no synchronized flowering and fruiting, which makes harvesting a labour intensive process.

Pongamia flowers once in a year and takes about 8 to 10 months to mature. Pods can be easily harvested/Collected after leaf shedding and before the new flush comes by shaking the branches. Rural women and unemployed youth can be gainfully employed for collection and will improve their livelihoods.

PROCESSING AND HANDLING

After collection the fruits are transported in open bags to the processing site. They are dried till all the fruits are opened and decorticated manually or by decorticators. It has been reported that direct sun drying has a negative effect on seed viability and the seeds should be dried in the shade.

YIELD POTENTIAL

Jatropha plant begins bearing from first year onwards and stabilizes from the fifth year. It gives an economic yield up to 30-40 years. The expected yield from one-hectrare plantation is near about 750-1500 kg under rainfed and 1500-2500 Kg under irrigated conditions. Pongamia starts bearing at the age of 7 years.

A single tree yields 10-200 Kg seed, indicating a minimum yield potential of 1000Kg/ha; however, grafted seedlings start bearing from the third year. Irregular bearing has been noticed in Pongamia and the yield may vary from year to year.

However, no systematic data on yield from block plantations are available. In Maharashtra, where block plantations on Jatropha were undertaken in the 1980s without irrigation, a yield of 1000-1200 Kg ha^1 was recorded.

There is an urgent need to study the yield potential of both the plants with appropriate agronomic management practices in block plantations.

STORAGE AND VIABILITY

The seeds should be dried to reduce the moisture content (5-7 per cent) and stored in cool and dry place. At room temperature the seeds can retain high viability for at least one year.

However, because of the high oil content the seeds cannot be expected to store for long time. Oil can be extracted using low cost expellers in the village itself and oil cake may be retained with the growers/collectors for application to fields as an organic fertilizer or can be sold to prospective buyers.

By-Product

Oil Cakes as Source of Plant Nutrients

Oil cake a by-Product after extraction of Oil appears to be a very attractive proposition as it contains all the macro and micro-nutrients and it is an excellent organic fertilizer unlike inorganic fertilizers that supply only few nutrients. Four kilograms of seed of Pongamia or Jatropha give about three kilograms of cake. The cake is mostly used for fertilizing plantation or commercial crops.

An analysis of the contents of the cake of Pongamia and Jatropha indicated the presence of all the essential elements required for plant growth, particularly found to be rich in nitrogen and sulphur. Recycling of Pongamia cake in adilabad district of Andhra Pradesh resulted in a 41-47 per cent increased income when compared to farmer's practice.

Table. Chemical Composition of Oilcakes Analysed at ICRISAT, Patancheru, India

Nutrients	Jatropha[1]	Pongamia[2]	Pongamia[3]
Nitrogen (%)	4.91	4.28	6.14
Phosphorous (%)	0.90	0.40	0.72
Potassium (%)	1.75	0.74	1.07
Calcium (%)	0.31	0.25	0.96
Magnesium (%)	0.68	0.17	0.35
Zinc (ppm)	55	59	95
Iron (ppm)	772	1000	1053
Copper (ppm)	22	22	41
Manganese (ppm)	85	74	108
Boron (ppm)	20	19	43
Sulphur (ppm)	2433	1894	3615

Table. Response of Soybean and Maize to the Application of Pongamia Cake and Inorganic Fertilizers

Crop	Treatment	Yield (kg ha^{-1})	Income (Rs. ha^{-1})	Cost of fertilizer (Rs.)	Additional income over farmer's practice (Rs. ha^{-1})
Soybean	Farmer's practice (DAP 75 kg ha^{-1})	900	10800	450	–
	Pongamia cake PC) (300 kg ha^{-1})	1340	16080	1500	4230
	Recommended dose of fertilizer (RDF) 20 kg N ha^{-1}	1450	17400	250	6800
	50% RDF + 50% PC	1650	19800	1500	7950
Maize	Farmer's practice (DAP 125 kg ha^{-1})	1200	6000	1125	–
	Pongamia cake (PC) (1800 kg ha^{-1})	2240	11200	4000	2325
	Recommended does of fertilizer (RDF) 90 kg N ha^{-1}	2390	11950	1000	6075
	50% RDF + 50% PC	2560	12800	5000	2925

Similar studies carried out by Ngoma in Zimbabwe have revealed that application of Jatoropha cake at the rate of 0, 0.25, 0.5 and 1.0 Kg/ sq. m resulted in cabbage yields of 16.8, 23.6, 22.8 and 35.8 Kg, respectively and the crop was free from pest and diseases. Tasosa *et. al.* have recorded a significant difference in the growth rates of tomato and the total above ground dry matter with increased applications rates of Jatropha and castor cakes. Henning is of the opinion that Jatropha plants can reduce the soil and water erosion when planted as live fence and the cake obtained after oil extraction can help in building the organic matter content of the soils of Sahelian countries.

Substitution of oilcakes with fertilizers is likely to improve the fertility of the soils in the long run and the soils will overcome the widely observed deficiency of several nutrients like N, P, Zinc, Boron, Sulphur, etc. Further, it will reduce the depencence of farmers on external input (fertilizers). Recycling of the cake serves the interest of farmers as well as of the government as the huge subsidy paid on fertilizers to industries can be reduced. The amount saved on subsidy of fertilizers can be used for encouraging biodiesel plantations in rural areas for ensuring energy, food and livelihood security.

Plantation Management-Strategies for Livelihoods Enhancement

In the semi-arid tropical region, large numbers of poor people reside and most of them are landless and marginal small landholders. Biodiesel plantations could be successfully utilized to improve livelihoods on a sustainable basis and to rehabilitate the degraded lands and protect/improve the environment. There is already a good market demand globally for biodiesel with spiraling fossil fuel prices and the increased awareness of the adverse impact of global warming. Good policy support from the national governments also exists, e.g. in India and Thailand where governments have extended good support for enhancing production and use of biodiesel.

Technology options to process biodiesel using multi-feed stocks also exist. However, the missing factor or the fourth wheel to run the biodiesel car is an assured supply of raw materials. There is an urgent need to enhance production and supply of Jatropha and Pongamia and other non-edible oils to be used as biodiesel. In densely populated countries of Asia as in Europe edible oils cannot be used for biodiesel production as already these countries are in the short supply of edible oils for human consumption. Similarly, to achieve oil security most of the developing countries would not like to sacrifice their food security. The proposed strategy is to bring vast areas of degraded and low quality lands under biodiesel plantations.

Collector's Model

In many Asian countries already a large number oil seed trees exist as avenue plantations or in forest areas. For example, in India Pongamia pinnata and Azadirachta indica trees are commonly grown as avenue plantations on common lands as well as on riverbanks. At present the seeds from these trees are not commercially used to the possible extent.

The production needs to be enhanced and this requires favourable strategies where growers are not put to hardship and loss. One of the options in promoting livelihoods of the poor is through the promotion of plantations by user groups or self-help-groups (SHGs) on common pool lands like degraded forests, community owned lands, low quality lands not suitable for annual food crops production, railway setbacks, canal embankments, tank foreshore, etc. The usufruct rights for harvests are ensured to the groups who manage the plantations but the right on land is not extended. Through assured market demand the landless people in groups (SHGs) could handle collection from the decentralized plantations and earn their livelihoods. This strategy along with the strategy to develop/ rehabilitate degraded lands through biodiesel plantations could ensure an efficient use of existing natural resources for promoting a win-win solution through biodiesel plantations.

Decentralized Extraction and Centralized Processing

There is a need to promote and utilize the expellers available locally as this will minimize the cost of transport of raw material, and will generate employment in rural areas. This will also ensure an availability of cake in the rural areas and the recycling of cake back to the fields. There is a possibility of using oil directly in the irrigation pumpsets and it can be blended directly with diesel up 10 per cent for use in farm machinery. Electricity generation is also possible in remote areas using filtered oil directly but all these require decentralized extraction. For esterification and biodiesel production a centralized processing facility can be created, which will minimize the cost of transporation and ensure strict quality adherence.

Recycling of Cake for Production of Biogas and Compost

The cake offers an excellent opportunity for reducing the dependence on fertilizers and its marketing will generate additional income. The cake can also be used for production of biogas before composting and both these processes will improve nutrients availability as fertilizer to the crops during the season itself as well as reduce pressure on energy demand for domestic use.

Carbon Trading by Women SHGs

The associated benefit with biodiesel plantations is small income for the rural communities through carbon trading. Under natural conditions Pongamia Pinnata was observed to sequester 17, 72, 331 and 347 Kg carbon per plant at 5, 10, 15, 25 years of age, respectively in a study carried out in Adilabad district. Adilabad district of Andhra Pradesh (AP) state in India is richly endowed with a good forest cover (47 Per Cent), black soils (Vertisols), and substantial rainfall (1100mm). The district has a large presence of indigenous people, comprising gonds, Lambadas and others.

Powerguda, a remote tribal hamlet in the Adilabad district, became an environmental pioneer when it sold the equivalent of 147 tons of carbon dioxide in verified emission reduction as carbon replacement to the world Bank in October 2003. The World Bank paid US $ 645 to Powerguda women SHGs to neutralize the emissions from air travel and local transport by international partidcipants attending its international conference in Washington, USA held on 19-21 October 2003. This was the first time that the multilateral agency made a direct payment to an Indian village for exporting environmental services. The emission reduction was calculated on the basis of 51 tons of Pongamia oil substituting for petroleum diesel over 10 years from the planting of 4500 Pongamia trees. The women's group are able to raise 20000 seedlings of Pongamia and Jatropha from this income every year and generate further income through sale of seedlings.

Most of seedlings are sold to the forest department, but some are also planted on field boundaries, farm bunds and community owned lands. The women are also members of the forest protection committee formed to protect the nearby forest under the government's Community forest management project. Other village activities include social networking, watershed management, improved agricultural practices and income generating activities. The establishment of an oil mill to crush Pongamia seeds into oil has helped the women to increase income through the sale of oil and oilcake.

Table. Carbn Calculations for Powerguda Village, 2003-2012

Year	Oil Yield (kg)	Trees	Total oil yield (kg)	C(t)	CO_2 eq (t)	Value (US$)	Discount value (at 3%)	NPV
2003		3,600	410	0.32	1.17	6.72	1.00	6.72
2004			494	0.39	1.41	8.09	0.97	7.85
2005			590	0.46	1.69	9.66	0.94	9.08
2006	0.5		1,125	0.88	3.22	18.43	0.91	16.77
2007	1		3,600	2.81	10.31	58.97	0.88	50.71
2008	1.5		5,400	4.21	15.46	88.45	0.85	51.89
2009	2		7,200	5.62	20.61	117.94	0.82	96.71

2010	2.5	9,000	7.20	26.43	151.24	0.79	119.48
2011	3	10,800	8.42	30.92	176.90	0.76	134.45
2012	3.5	12,600	9.83	36.07	206.39	0.73	150.66
		51,219	40.13	147.29	842.79		644.32

The pioneering work of extracting oil from Pongamia seeds, exporting environmental services to the World Bank and improved income level has given the people a sense of pride and has put the village on the map of the world. Women's involvement in nursery raising, plantation, carbon trading has triggered the social capital development and social cohesion.

Today, the women are able to manage their finances, are sensitized about health and nutritional issues such as HIV, etc. and need education for all. There is no migration from the village and this may be attributed to an increase in the level of income in each family.

The improvement has come mainly from agriculture (93 per cent) and forest (7 per cent) which translates to Rs. 27,821 (2002-03) from Rs. 15677 (1999-2000). In 2003, the four groups of SHGs in Powerguda had Rs. 552000/- as total savings, which worked out Rs. 6608/- per household.

Now, they are in a position to get loans from banks and are out of the clutches of moneylenders. Powerguda action has inspired several neighboring village to plant Pongamia/Jatropha trees on a large scale and this has been repeated in 20 other villages in Adilabad district.

Returns from Plantations

Not much research work has been undertaken on these plantations and their economic returns. Several types of returns from the growing and the use of products from Jatropha and Pongamia need to be carefully estimated. There is a need for research on the operational scale to identify plus trees and improved management techniques for higher productivity and economic viability. The economic returns may vary from place to place depending on the climate, soil and management practices.

The tangible benefits, include the market prices of different products like diesel and by-products like cake and glycerol while intangible benefits are many. Besides tangible benefits, there are innumerable intangible benefits like:

- Potential to produce a green fuel that will reduce carbon dioxide emissions as biodiesel will be recirculating atmospheric CO_2 through the process of photosynthesis.
- Greening of waste and marginal lands, which has positive impact on hydrology.

- Alleviate soil degradation, desertification and deforestation and conserve soil and rainwater.
- Improved ecosystem and environmental sustainability.

Tree based oilseeds offers scope for rehabilitating degraded lands and improve livelihoods in developing countries. Green cover over barren and unproductive lands will reverse the process of degradation caused mostly by water erosion. The fertility of these marginal lands will improve through recycling of nutrients from the deeper layers, addition of leaf litter, nitrogen fixation (Pongamia) and carbon will be sequestered unlike fossil fuels.

Studies have shown that Jatropha returns 19 Kg N ha^{-1} $year^{-1}$ through litter fall from the third year onwards. Employment generation from plantation, harvest and processing activities will reduce migration from the rural areas, which is a big concern in most Asian countries. Participation of women SHGs in managing the plantations will boost their livelihoods and will empower the women.

Recycling of cake will reduce the dependence on inorganic fertilizers and will pave the way for organic farming. Further, there is scope of earning carbon credits and the additional benefits can be passed on to the growers.

As necessary market demand, government support and technologies to process do exist, however, the potential cannot be tapped mainly because of insufficient availability of feedstock. Biodiesel plantations for rehabilitating degraded lands, and increasing income from low quality lands provides a win-win situation for protecting the environment, increasing income, and generating employment opportunities in rural areas.

However, the lack of scientific information about suitable management practices as well as identification of high-yielding accessions and lack of high yielding cultivars on a large scale holds back development in this area. There is an urgent need to undertake detailed research for development of management practices, high-yielding cultivars, mass multiplication techniques, suitable development models for increasing supply of raw material for biodiesel production.

The by-product cake after extraction of oil is a potential source of valuable plant nutrients as well as organic matter for improving the soil quality of tropical soils, and is a good source of energy through biogas for meeting the domestic energy needs. In conclusion, the development of biodiesel plantation provides win-win situation and needs targeted investments towards research and development aspects for ensuring the supply of raw material.

Chapter 10

Reduction in Organochlorine Pollutants

The Palo Alto Regional Water Quality Control Plant (RWQCP) releases various organochlorine pesticides, including chlordane, DDT, dieldrin, heptachlor, and toxaphene, to the San Francisco Bay through its treated effluent and could potentially release them to soil if sludge management practices change.

RWQCP discharges are of concern because of the human and environmental toxicity of organochlorine pesticides and their environmental persistence. As a result, the RWQCP seeks to reduce the amount released by reducing the sources of certain organochlorine pesticides in the plant's influent. The sources and conveyances of organochlorine pesticides to the wastewater treatment plant were identified in Organochlorine Pesticides Source Identification.

Using the sources identified and the recommended source reduction strategies in that report, a pollution prevention plan was developed for the large conveyances of these pesticides to the RWQCP. This document is intended to supplement the RWQCP's primary pollution prevention plan, the Clean Bay Plan. The RWQCP intends to include the control strategies identified in this report as "strategies for immediate implementation" in the Clean Bay Plan 1998.

During the RWQCP's annual pollution prevention planning process, the RWQCP will also consider implementing actions identified in this report as "strategies for further consideration." During the RWQCP's annual planning process, the RWQCP will evaluate any additional pollution prevention opportunities identified through follow-up studies and other staff activities for inclusion in the Clean Bay Plan.

SOURCES OF ORGANOCHLORINE PESTICIDES

The use of all of the organochlorine pesticides of concern for the RWQCP has been banned in the United States. The only exception is heptachlor, which, while banned in California, still may be used in other parts of the U.S. for control of fire ants at utility installations through subsurface injection. The pesticides identified in this report continue to be used in much of the world, particularly in developing countries. Their persistence in the environment allows the pesticides to circulate through and remain in the environment as a result of historical and current worldwide uses.

As a result, air deposition of pesticides, pesticides applied to foods grown in other parts of the world and materials manufactured elsewhere and imported into the United States, pesticide residues on previously treated surfaces, and pesticides accumulated in the food web are all sources of organochlorine pesticide discharges to the RWQCP. Although organochlorine pesticides from these sources are not directly discharged to the RWQCP, a variety of conveyances can transport the organochlorine pesticides from these sources to the RWQCP. In general, the RWQCP may have the ability to control the conveyance of organochlorine pesticides to the RWQCP, but will have little control over the actual source.

In Organochlorine Pesticides Source Identification, conveyances of organochlorine pesticides were classified as either relatively large or relatively small contributors to the RWQCP's load. Table lists the large conveyances to the RWQCP and the estimated loads from these sources. Because very limited data exist on organochlorine pesticide discharges, these estimates have very high uncertainty. Because of the uncertainty of the estimated loads, the individual results should not be compared to each other or added together to estimate the total contribution of the known organochlorine pesticide conveyances.

As described in the source identification report, the conveyances are described as either large or small contributors to the RWQCP's load to avoid these comparisons. Source specific control strategies were developed for the large conveyances, and are presented in the following pollution prevention plan.

Table. Large Conveyances to the RWQCP

Large Conveyances to RWQCP	Pesticides with Potentially Large Loads from the Conveyance	Estimated Load (10^4 lb/day)	Most Important Sources
Storm Water Inflow	Chlordane	<10	• Air deposition of pesticides circulating in the enivronment
	DDT	<20	
	Dieldrin	<4	
	Heptachlor	<70	• Pesticide residues on

	Toxaphene	<7	treated materials and soils • Pesticides applied to imported materials
Human Waste	DDT	10	• Organochlorine pesticides in food web
	Heptachlor	1	• Pesticides applied to imported foods
Food Waste	DDT	5	• Organochlorine pesticides in food web
	Dieldrin	4	
	Heptachlor	3	• Pesticides applied to imported foods
Water Supply	Dieldrin	<40	• Air deposition of pesticides circulating in the environment
Total Estimated Effluent Load	Chlordane	7	
	DDT	2	
	Dieldrin	4	
	Heptachlor	0.2	
	Toxaphene	20	

ORGANOCHLORINE PESTICIDES POLLUTION PREVENTION PLAN

The pollution prevention plan for organochlorine pesticides identifies strategies for controlling levels of organochlorine pesticides from the large conveyances to the RWQCP, including storm water inflow, food wastes, and human wastes.

For each conveyance, a short description of the sources of organochlorine pesticides to the conveyance is followed by the control strategies that the RWQCP plans to implement or is considering implementing in the future in an effort to reduce the influent organochlorine pesticide load to the plant. The following section describes the general categories of control strategies that could be implemented.

GENERAL CONTROL STRATEGIES

Source control strategies generally fall into four different categories:

- Technology-based Strategies.
- Local Regulatory Strategies.
- Public and Business Outreach.
- Regional, National, or International Strategies.

For each of these general categories, a short description of the elements of a general strategy and examples of programme effectiveness are presented below.

TECHNOLOGY-BASED STRATEGIES

Technology-based strategies involve a process modification or the use of equipment or chemicals to achieve reductions in organochlorine pesticide discharges. The effectiveness of these strategies depends on the ability of the technology to remove organochlorine pesticides from the source or the organochlorine pesticide discharge from the sewer as well as the amount of cooperation and participation that is necessary to implement the strategy.

LOCAL REGULATORY STRATEGIES

Local regulatory strategies include local ordinances and controls on specific industries or sources. These controls may include product bans or restrictions on a local level or enforced discharge limits and site visits for businesses. The feasibility of this strategy is limited by state and federal law (e.g. FIFRA prohibits many local restrictions on pesticides) and by the ability of dischargers to control the source of organochlorine pesticides in their discharges. The effectiveness of this type of strategy depends on the ability of the RWQCP to enforce the regulations.

PUBLIC AND BUSINESS OUTREACH

Regulatory strategies are generally not applicable in the residential sector and technology-based strategies are difficult to implement on a large, residential scale. Therefore, public education and outreach are often the most effective ways of implementing source reduction in the residential sector. Outreach and education to businesses can be a cost-effective way of raising awareness and changing business habits for source control without the difficulties and time involved in implementing legislative or technology-based controls.

Public and business outreach can be accomplished in many different ways. Brochures, point of purchase (POP) displays, event displays and media advertisements all can present clear, concise information to a broad audience. School programs can be used to create awareness in children which they can then pass on to their parents. Business outreach can take the form of education on alternative products and best management practices.

Business incentive programs in which businesses are recognized for following pollution prevention guidelines or providing educational materials to their customers are effective ways of educating both business owners and the public about environmental issues.

REGIONAL, NATIONAL, OR INTERNATIONAL STRATEGIES

Many of the organochlorine pesticide sources are not easily controllable by the RWQCP, but the RWQCP can approach other entities to advocate the reduction of organochlorine pesticides from these sources. Strategies falling into this category may include working with regional, national, or international groups to create organochlorine pesticide reduction strategies.

SOURCE SPECIFIC CONTROL STRATEGIES

This plan presents a menu of potential strategies for reducing organochlorine pesticide discharges to the RWQCP. The document divides these control strategies into two categories: strategies for immediate implementation, and strategies for further consideration. The RWQCP plans to initiate implementation of the "strategies for immediate implementation" in the near term, because these strategies have relatively low barriers to implementation and are currently feasible. The RWQCP plans to evaluate the "strategies for further consideration" during its annual planning process and when conditions affecting these strategies change, with the potential for implementation in the future.

STRATEGIES FOR IMMEDIATE IMPLEMENTATION

Residential Pesticide Collection Event

Although organochlorine pesticides have been banned for use in California for several years, it is possible that some residents still have leftover amounts of these pesticides. In association with existing household hazardous waste collection programs, the RWQCP could encourage residents to turn in all their organochlorine pesticide products. The RWQCP could provide information to residents that identifies pesticides that have been banned for use in California and encourage them to turn the pesticides in. The information provided should include why the pesticides should not be disposed of in the trash or down storm drains or sewers. This programme should be coordinated with local household hazardous waste management personnel.

STORM WATER INFLOW

Organochlorine pesticides reach storm water from a variety of sources. Organochlorine pesticides circulating in the environment can deposit on surfaces that contact storm water. Outdoor surfaces that historically were treated with organochlorine pesticides may contain pesticide residues that come into contact with storm water. Since these pesticides are still used in other parts of the world, pesticides could be applied to materials grown and manufactured elsewhere and imported into the United States.

The pesticide residues on these materials could come in contact with storm water. Although these pesticides are not being used in California, their persistence in the environment means that uses in other parts of the world could contaminate storm water runoff in the RWQCP's service area. The RWQCP could notify appropriate agencies about this issue to raise their awareness and encourage a reduction in the use of organochlorine pesticides worldwide and the resulting environmental circulation of these pesticides.

STRATEGIES FOR FURTHER CONSIDERATION

The following strategies involve diversion of waste streams from the RWQCP to land in the RWQCP service area. Because these strategies do not prevent the release of organochlorine pesticides, they are less desirable than strategies that address the ultimate source of organochlorine pesticide release to the environment. While pollutant transfer to land is less desirable than source reduction, the diversions discussed below has environmental merit in that they involve reuse of a resource (compost), rather than disposal of the material as a waste.

HUMAN WASTE AND FOOD WASTE

Organochlorine pesticides in human waste result primarily from pesticides in foods ingested by residents. Air deposition onto soil and water and bioaccumulation in the food web leads to contamination in foods. While the RWQCP has no control over organochlorine pesticide levels in food, it does have some control over the pathways of food waste to the sewer system. Food waste ends up at the RWQCP after residents put left over food down their disposals.

Educating the public about alternative disposal methods, such as composting, would reduce the amount of food waste reaching the RWQCP by diverting the food waste and its associated organochlorine pesticides to individual gardens. The RWQCP could institute an education campaign for residents on how to compost household food waste, sources of information on composting, and resources for disposing of any excess compost material.

The RWQCP could support community composting workshops that teach the techniques of composting and hand out education materials. Such an outreach programme should be conducted in conjunction with local solid waste programs that encourage composting. The RWQCP should consider encouraging the use of kitchen storage containers or in-kitchen worm boxes as an element of this programme. Diverting food waste from the sewer system could affect wastewater treatment operations at the RWQCP.

Investigation of potential operational consequences would need to be conducted prior to initiation of any large-scale programme in this area. Composting toilets have been proposed as a possibility for controlling contamination from human waste. However, composting toilets do not currently comply with the plumbing code in Palo Alto. Substantial research would have to be conducted before installation of composting toilets in the RWQCP service area could be considered.

ESTIMATED LOAD REDUCTIONS

Due to the uncertainty associated with the sources of organochlorine pesticides to the RWQCP, it is very difficult to estimate quantitatively the potential load reductions from the control strategies presented in this plan. Assuming that a load reduction is possible and needed information has been collected, a general assessment of possible load reductions. The numbers presented are based on very uncertain estimates of possible reductions and the resulting load reductions could largely over or underestimate the possible reductions.

PARTICIPATION FACTOR

Ideally, the implementation of a control strategy would result in the elimination of the source it was designed to address. In reality, only a certain percentage of the people and procedures addressed by the control strategy will be changed. The participation factor estimates the amount of participation that can be achieved for a given control strategy. Participation factors used in this study are described below. Because limited data on participation levels are available, these participation factors should be considered to be rough estimates.

Whenever possible, estimates from previous efforts in the Palo Alto area are used. Public education and outreach programs can be expected to have an effectiveness between 5 and 20%. Outreach efforts in Seattle resulted in behaviour changes in 6-13% of the people surveyed. In Palo Alto, the return rate for car wash coupons mailed to residents as an outreach measure was 10% in 1995 and 9% in 1996.

However, since many of the outreach messages for controlling sources of organochlorine pesticides are more complex than the car wash campaign, a 5% participation factor is used for the public education programs. The participation rate for regional/national/international strategies to reduce worldwide organochlorine pesticide use is very difficult to estimate because of a lack of information about who would be involved in implementing this strategy and its technical feasibility.

Since a significant part of the programme would probably involve educating regional and national entities on the connection between organochlorine pesticide use and stormwater quality, the strategy was assumed to have the same participation rate as a difficult to implement public education campaign (5%).

LOADING FACTOR

The loading factor is the amount of organochlorine pesticides load reduction from a source that could be expected if there was 100% participation. The loading factor varies depending on the sources that the strategy addresses. For example, all programs related to food waste have a loading factor of 100%. On the other hand, programs for storm water inflow may only address the sources from spills or air deposition and therefore have a lower loading factor.

Loading factors were determined by estimating the amount of organochlorine pesticides coming from individual sources within a category. For storm water inflow, it was assumed that the majority of the pesticides in the inflow would come from air deposition and sources outside the United States. However, the proposed strategies could only address future releases of the pesticides to the environment, not the amount already circulating in the environment. Therefore, a loading factor of 10% was used for the international strategy to reduce the use of organochlorine pesticides around the world.

Since many of the sources of organochlorine pesticides to the RWQCP are a result of the environmental levels of organochlorine pesticides, any actions that the RWQCP takes to reduce the amount of organochlorine pesticides entering the environment may reduce the load to the plant. In an attempt to provide a quantitative evaluation of the source reduction strategies, the assumed participation and loading factors and the estimated load from each source presented in Organochlorine Pesticides Source Identification were used to obtain an estimated load reduction.

As described in that report, the estimated loads are very uncertain as is the load to the RWQCP. Each of the source load estimates is subject to sufficient uncertainty that the estimates should not be added together to estimate the total contribution of the known organochlorine pesticide conveyances. In order to evaluate the results that could potentially be achieved through implementing the identified control strategies, the report categorizes potential reductions as large (reduction potential equal to or greater than 10% or the RWQCP influent organochlorine pesticide load), small (reduction potential between 1% and 10% or the RWQCP influent organochlorine pesticide load), or very small (reduction potential less than 1% of the RWQCP influent organochlorine pesticide load).

By categorizing both sources and potential reductions as large or small, this report avoids inappropriately comparing the individual results with one another. The pesticide collection strategy was not evaluated due to the lack of information about the amount of the load that potentially could be attributed to this source. Table shows participation and loading factors used to estimate reductions and the resulting load reductions for each strategy.

Human waste, air deposition from organochlorine pesticides circulating in the environment, historically treated materials and soils, and pesticides applied to imported materials and food are not easily controllable by the RWQCP. For this reason, control strategies were not developed for these sources. If additional information is found about these sources that changes their controllability, the RWQCP will address them in the future. The pollution prevention plan for organochlorine pesticides considers all of the large conveyances of organochlorine pesticides to the RWQCP.

For the most part, additional information on sources needs to be obtained before further actions can be taken. The control strategies that have been identified in this report have the potential to reduce the organchlorine pesticide influent load to the RWQCP by a relatively small percentage. As information is gathered, the RWQCP will determine whether any other control strategies can be developed that may result in additional load reductions.

Due to the uncertainty and lack of information available on the majority of the organochlorine pesticide sources to the RWQCP, the quantitative load reductions developed for this pollution prevention plan could significantly under or over represent the potential load reductions. Depending on the information found through the implementation of this plan, load reductions may be able to be achieved. However, the results are highly dependent on the quality and quantity of information obtained and the magnitude of the sources for which control strategies can be employed. As described in the Organochlorine Pesticides Source Identification report, organochlorine pesticides are persistent in the environment and disperse widely throughout the world. As a result, many of the pesticides reaching the

RWQCP come from this environmental dispersion, and the magnitude of the concentrations in the plant's influent is related to these environmental organochlorine pesticide levels. Recent studies have shown that these environmental levels of organochlorine pesticides are gradually declining.

This trend is expected to contribute to reductions in organochlorine pesticide levels in the RWQCP influent. Efforts made by the RWQCP to reduce organochlorine pesticide levels in its releases could help further this trend. The Organochlorine Pesticides Source Identification report recommends further studies be conducted to obtain additional information about several pesticide sources (notably, the water supply, storm water, and sediments in building sewer lines and laterals). The lack of information about these sources precludes development of control strategies for them.

PERSISTENT ORGANOCHLORINE POLLUTANTS

THE MARINE MAMMAL HEALTH

Through an agreement with the National Marine Fisheries Service (NMFS), Office of Protected Resources, NIST provides a service to the Marine Mammal Health and Stranding Response Programme (MMHSRP) by maintaining the National Marine Mammal Tissue Bank (NMMTB) as part of the National Biomonitoring Specimen Bank (NBSB). NIST personnel also develop and test collection and banking protocols, and provide training to other individuals who collect specimens for the bank.

One source of specimens for the NMMTB is the mass stranding of cetaceans. These events periodically occur along the coasts of the United States, particularly on the East Coast. Regional Stranding Networks consist of local and regional groups from federal, state, and local agencies; private organizations; and individuals, many of whom are volunteers. These people respond to mass strandings by providing aid to the live-stranded animals, returning the live-stranded animals to the ocean (when and where appropriate), and transporting animals to recovery facilities.

They also collect data and samples from animals that die to document the age/sex population structure, life history, health condition, and other information that might be useful for scientific purposes. The stranding network members, when trained appropriately, also collect specimens for contaminant analysis and for the NMMTB. NIST provides training, sample collection, and banking of marine mammal specimens for the MMHSRP and NMMTB.

Additionally, NIST provides retrospective and baseline analysis of organic and inorganic contaminants on selected marine mammal specimens from the bank as part of the National Marine Analytical Quality Assurance Programme. Additional details on NIST's role in the NMMTB, MMHSRP, and the National Marine Analytical Quality Assurance Programme are provided in Becker *et al.*

Persistent Organochlorine Pollutants and Trace Elements in Marine Mammals

Contamination of food webs by anthropogenic persistent organochlorine pollutants (POPs), such as polychlorinated biphenyls (PCBs), DDT, toxaphene, and chlordane, was first documented by Jensen and of marine mammals by Holden and Mardsen. Subsequently, researchers have determined that the release of POPs during their manufacture, storage, and use has resulted in global contamination, particularly of the marine environment. Due to their persistence and highly lipophilic nature, many POPs bioaccumulate and biomagnify in the food chain. POP residues have been detected in all levels of aquatic biota ranging from phytoplankton to marine mammals. Biological effects of these compounds on mammals include hepatocellular carcinoma; reproductive, endocrine, and developmental toxicity; immunotoxicity, and wasting syndrome.

Toothed whales (odontocetes) are particularly susceptible to bioaccumulation and the effects of POPs due to many factors including their relatively high trophic position, long life spans, limited metabolic capability to degrade chemical contaminants, and reproductive strategies. For many marine mammal species, the ranges of normal elemental concentrations are not known. Levels of some metals in marine mammal tissues are often higher than concentrations found in terrestrial mammals (i.e., mercury and cadmium) and there are often marked differences in element levels among different marine mammal species. There also may be differences among animals of the same species but from different locations. Concentrations for selected elements in various tissues have been reported for common dolphins (Delphinus delphinus), harbor porpoise (Phocoena phocoena), pilot whale (Globicephala melas), beaked whales (Ziphiidae) and many other species of odontocetes or "toothed whales", but there are no trace element data for rough-toothed dolphins (Steno bredanensis).

THE ROUGH-TOOTHED DOLPHIN

Rough-toothed dolphins are pelagic odontocetes that are widely distributed in tropical and warm temperate oceans around the world.

Most information on this species has come from populations in the eastern Pacific Ocean, with little available information on the populations inhabiting the Gulf of Mexico or western Atlantic Ocean. Growth occurs most rapidly during the first five years of life. Both sexes attain maximum lengths of approximately 280 cm with males reaching sexual maturity at approximately 225 cm or 14 years of age, and females at 210 cm or 10 years of age. Morphological features that distinguish this species from other members of the delphinid family (e.g., bottlenose dolphin (Tursiops truncatus)) include the lack of visible demarcation between beak and melon, and the rugose tooth surface from which its common name is derived.

Coloration varies geographically, but animals are generally dark gray dorsally with variable pinkish and white markings ventrally. Rough-toothed dolphins are social animals, typically found in small groups ranging from 10 to 20 individuals with larger groups reported. Their diet mainly consists of pelagic fish and squid. The authors are aware of only two previous publications describing organic contaminants in this species-Marsili and Focardi (1997, $n = 1$) and O'Shea *et al.* (1980, $n = 7$). The objective of this report is to document POP and inorganic constituents in tissues of rough-toothed dolphins from the Gulf of Mexico. Prior to this report there were no data on POPs or inorganic constituents in this species from this region. This report contains both organic and inorganic data, generated by NIST, for dissemination to sponsors and user groups. These data will help to establish baseline levels of POPs and both essential and potentially toxic elements in the tissues of this species.

SPECIMEN COLLECTION

On December 15, 1997, 62 rough-toothed dolphins stranded on the Gulf Coast of Florida near Apalachicola at St. Joseph State Park (29°45.2′ N, 85°25′ W). All the animals were pushed back out to sea. However, 34 animals restranded and of those, 17 died on site, 1 swam away, and 16 were sent to rehabilitation facilities, of which 4 were released.

Fifteen of the restranded animals that died on the beach were necropsied and sampled for the NMMTB. In addition to the tissues sampled for the NMMTB (blubber, kidney, and liver) and real-time contaminant monitoring (blubber, melon, and muscle, NOAA/NMFS/ Northwest Fisheries Science Centre (NWFSC)), samples were provided to the Armed Forces Institute of Pathology for histological and pathological analyses. Teeth were also removed from the animals for age determination. Samples for the NMMTB were collected and processed using standard NIST protocols. These protocols consist of the removal of tissue specimens from each animal using standard procedures and equipment designed to minimize the contamination of the tissues during removal and handling. Because of the rapid deterioration of tissues following death, it was decided that only animals that strand alive and die of natural causes or are euthanized, or animals that are considered "freshly dead," are sampled.

Materials that contacted the specimens were limited to Teflon and titanium. The tissue samples (blubber, kidney and liver) were excised from the animals using titanium knives, placed in Teflon bags, and transported on ice to the NOAA/NMFS Panama City Laboratory. At the laboratory, each specimen was divided into Samples A and B (approximately 150 g, each) according to the NMMTB protocols, placed in pre-cleaned Teflon sample jars, frozen, and express shipped to the NBSB, NIST Gaithersburg, MD in liquid nitrogen vapour shippers. Upon arrival at NIST, samples were immediately placed in liquid nitrogen-vapour cooled (– 150 °C) freezers.

Tissue Sample Preparation

Each tissue specimen to be analysed (approximately 150 g, each) was homogenized using a cryogenic procedure designed to reduce the likelihood of changes in sample composition due to thawing and refreezing. Subsamples of the tissue homogenate, a frozen (non freeze-dried) powder, were transferred to Teflon jars (10 mL) for storage (at – 150 °C) until analyses were performed. Portions used for instrumental neutron activation analysis (INAA) were lyophilized prior to analysis.

Age Determination

Teeth were extracted from the jaws during necropsy or after maceration in the lab. Teeth were sectioned longitudinally to a thickness of 2 mm using a low-speed saw. Sections were then placed in 10% neutral buffered formalin for approximately 24 h. Samples were removed and rinsed in running tap water for approximately 3 h and then soaked in tap water for at least 24 h. Sections were placed in approximately 80 mL of a decalcifying agent. Decalcification was prolonged (relative to Tursiops teeth) and took approximately 6 h to 16 h for completion.

Decalcified sections were rinsed in running water for 6 h and then soaked in tap water for at least 24 h. Teeth were serially sectioned using a sledge-type microtome with a freezing attachment. Each 2 mm section was further sectioned to 25 ìm then soaked in tap water for at least 1 h to remove any additional acid. Thin sections were stained in Mayer's hematoxylin for 75 min, rinsed in tap water for 1 min, blued in 0.5% ammonium hydroxide, and then rinsed again. Teeth were placed in 50% glycerin for 30 min and then placed in 100% glycerin for 24 h to complete the exchange of water to glycerin.

Sections were mounted on slides in 100% glycerin. This method has been used successfully for other small delphinids. Sections were examined under 10X to 60X magnification using a stereoscopic microscope. Growth layers were generally read using guidelines developed for Tursiops by Hohn *et al.* Each growth layer group (GLG) consisted of a dark and light staining layer.

Sections were read three times in the blind and the average reading was considered the final estimate. In a few cases, when readings were several layers apart, a fourth reading was taken and a subjective decision was made for the final estimation. Several of the teeth had extensive damage caused by extraction and could only provide minimum age estimates. For quality assurance purposes, the process was repeated using a second and sometimes third tooth (when available).

In such cases, all sections were evaluated and readings were made from the best-mounted sections. This procedure was necessary for most of the older animals. Ages were estimated to the last fully formed GLG except in younger animals. Animals with less than 8 GLGs were read to the nearest 0.5 GLG and animals <1 GLG were aged to the nearest 0.25 GLG. It has been established with other small delphinids that one GLG corresponds to 1 y of life. This was assumed to be the case for rough-toothed dolphins, although this relationship has not been verified directly.

Analysis of Blubber for Persistent Organochlorine Pollutants by NIST Charleston

Cryo-homogenized blubber samples were shipped from the NMMTB in Gaithersburg, MD to the NIST Charleston Laboratory in a liquid-nitrogen-vapour shipper. Upon arrival specimens were stored at –80 oC until required for analysis. The Teflon jar containing the sample was removed from the –80 oC freezer and immediately placed in a small Dewar flask containing approximately 100 mL of liquid nitrogen. Between 0.792 g and 1.21 g of sample was removed from the jar using a clean metal spatula that had been immersed in liquid nitrogen to prevent the frozen sample from adhering to the spatula.

The frozen sample was placed in a clean 150 mL beaker and weighed to the nearest 0.001 g. A small Styrofoam block was placed on the balance pan under the beaker to provide insulation and prevent balance drift. The sample was mixed with 30 g of sodium sulfate (Na_2SO_4; combusted at 700 °C for 24 h, then cooled in a desiccator prior to use). The mixture was then transferred to a 33 mL pressurized fluid extractor cell to which 10 g of Na_2SO_4 had been previously added.

A small volume (approximately 1 mL) of dichloromethane (CH_2Cl_2) was added to rinse the beaker, and then transferred to the PFE cell. For quality assurance purposes, RTDL-004 was analysed independently three times, and two blanks and two aliquots of NIST SRM 1945 Organics in Whale Blubber were prepared and analysed. Each blank consisted of approximately 45 g of Na_2SO_4 in a PFE cell. The samples of SRM 1945 were prepared in the same manner as the dolphin blubber samples.

A mixed internal standard solution containing 4,4′-DDT-d8, 4,4′-DDE-d8, 4,4′-DDD-d8, endosulfan I-d4, PCB 103, and PCB 198 was then added to the PFE cells, by weighing (to the nearest 0.00001 g) approximately 1 mL of the solution. Six calibration solutions were prepared by weighing (to the nearest 0.00001 g) portions of SRMs 2261 Chlorinated Pesticides in Hexane, 2262 Chlorinated Biphenyl Congeners in 2,2,4-trimethylpentane, 2274 PCB Congener Solution-II in Isooctane and 2275 Chlorinated Pesticide Solution-II in Isooctane into weighed portions of isooctane. The calibration curve ranged from approximately 5 ng to 1000 ng added to the PFE cell.

Approximately 1 mL of the appropriate mixed calibration solution was weighed (to the nearest 0.00001 g) in an HPLC syringe, then added on top of the Na_2SO_4. The mixed internal standard solution was then added to each PFE vial, by weighing (to the nearest 0.00001 g), approximately 1 mL of the solution. The samples were extracted with CH_2Cl_2 using PFE. The conditions were as follows: the cell temperature was 100 °C, equilibration 5 min, static time 5 min, cell pressure was 6.89 MPa and there were three cycles (one-third of the solvent each time).

The sample extracts were then reduced to between 0.5 mL to 1 mL by evaporation in a stream of purified nitrogen using a Turbovap II. High-molecular-mass compounds were removed by size exclusion chromatography (SEC) using a 600 mm x 25 mm (10 im particle size with 100 Å diameter pores) PLGel column as described in Kucklick *et al.* The extract was then fractionated using a semi-preparative aminopropylsilane column into relatively lower- and higher-polarity fractions (F_1 and F_2, respectively).

F_1 consisted of 50 mL of hexane, while F_2 consisted of 60 mL of 25% CH_2Cl_2 in hexane. Target compounds contained in F_1 included PCBs, heptachlor, oxychlordane, 2,4′-DDE, 4,4′-DDE, 2,4′-DDT, hexachlorobenzene (HCB), and mirex. Target analytes in F_2 included 4,4′-DDT, cis- and trans-chlordane, cis- and trans-nonachlor, α-, β-, and γ-hexachlorocyclohexane (HCH), heptachlor epoxide, 2,4′- and 4,4′-DDD, and dieldrin. Organochlorine compounds were determined by injecting each sample twice into a gas chromatograph (GC) with dual micro-electron capture detectors (ECD).

Organochlorines in F_1 and F_2 from the aminopropylsilane column were separated using a 60 m 5% phenyl methylpolysiloxane column with 0.25 mm internal diameter and a 0.25 ìm film thickness and a 60 m DB-XLB (proprietary phase, J&W Scientific) with 0.25 mm interior diameter and a 0.25 ìm film thickness. The instrument was configured by installing a 5 m x 0.25 mm retention gap to the inlet, then attaching a glass Y connector to the free end.

The columns were connected to the Y splitter, then to the ECDs. The injector and detector temperatures were 220 °C and 325 °C, respectively; the carrier and makeup gasses were H_2 (constant velocity of 30 cm/s) and N_2 (60 mL/min), respectively. Samples were injected into the GC (2 µL, splitless injection), and the oven was programmed from 90 °C initially (1 min hold) to 170 °C at 25 °C/min, then 1 °C/min to 260 °C, then ramped to 300 °C at 15 °C/min (10 min hold, 107 min total run time).

The amount of each compound in the unknown was calculated using the mass of internal standard added and the slope and intercept of either the entire six-point calibration curve generated from the response of the calibrants or those generated from the three calibrants which bracketed the concentration. Different internal standards were used to quantify target analytes to avoid suspected interferences and to maximize repeatability and reproducibility.

PCB 103 was used to quantify all F_1 analytes other than 4,4′-DDE, which was quantified using 4,4′-DDE-d8. All F_2 analytes were quantified using endosulfan I-d_4. Analyses of most F_1 target analytes were made with the DB-XLB column, as it gave baseline resolution on the majority of PCBs, and using PCB 103 as the internal standard. PCBs 87 and 101 were quantified with the DB-5 column, using PCB 103 as the internal standard, because of coelution problems on the DB-XLB column.

All F^2 target analytes were quantified using GC-ECD with the DB-5 column and a splitless injection. Detection limits were calculated as three times the concentration of a given compound in the mean blank, which gave more conservative detection limits than using the mean blank plus three standard deviations. The total non-volatile solvent extractable material was measured on subsamples (50% of total) taken from the original extracts by first reducing the solvent volume from the extraction with the Turbovap, then allowing the remaining solvent to evaporate at room temperature.

The extracts were repeatedly weighed until a stable mass was achieved. The ratio of the residue to the wet mass extracted represents the fraction of the total non-volatile solvent extractable material.

Analysis of Blubber for Persistent Organochlorine Pollutants by NIST Gaithersburg

The NIST Gaithersburg laboratory also analysed six rough-toothed dolphin blubber samples: RTDL-001 to RTDL-006. The sample extraction and clean up procedures were similar to those detailed above. Briefly, the blubber homogenates were mixed with Na_2SO_4 then added to PFE cells. The cells were spiked with an internal standard mixture containing PCB 103, 198, and perdeuterated 4,4′-DDT. The sample and calibrants were extracted and high-molecular-weight materials were removed from the extracts using SEC.

The chlorinated extract was reduced in volume then further isolated using a silica solid-phase extraction column (plus size, Waters Corporation, Milford, MA) that was precleaned using 15 mL of 10% dichloromethane in hexane. The fraction of interest was eluted also with 10% volume fraction of CH_2Cl_2 in hexane. The eluant was concentrated to approximately 0.5 mL for analysis. The concentrated samples were analysed using gas chromatography-mass spectrometry.

A 0.25 mm × 30 m fused-silica capillary column containing 5% phenyl-substituted methylpolysiloxane phase (HP-5ms, Hewlett Packard), 0.25 µm film thickness was used. The column was held isothermally (60 °C) for 1 min, temperature programmed at 45 °C per min to 180 °C for 30 min, and then temperature programmed at 2 °C per min to 250 °C where it was held isothermally for 15 min. All injections were 2 µL in a pulsed splitless mode. Helium was used as a carrier gas at a constant flow rate of 1 mL per min.

The MS transfer line was held at 280 °C. The major ions included 222, 255.95, 289 90, 323.9, 359.85, 393.80, 425.75, 463.75, 497.70 for the PCBs and 284, 181, 389, 353, 246, 371, 409, 235, 243, 195, 263, 272 for the pesticides. Seven-point calibration response curves for the analytes relative to the internal standards were determined by processing gravimetrically-diluted PCB and pesticide calibrant solutions. The per cent nonvolatile extractable material (mainly lipid) was determined for each sample after extraction.

The ëxtract was evaporatively concentrated to approximately 20 mL (mass known), and an aliquot of 90 ìL (mass known) was placed on an aluminum pan. The extract on the pan was air dried, and the mass of the dried extract was noted. The ratio of the residue to the wet mass extracted represents the fraction of the total non-volatile solvent extractable material.

ANALYSIS OF LIVER AND KIDNEY FOR TRACE ELEMENTS

Instrumental Neutron Activation Analysis (INAA) for 35 Elements in Livers and Kidneys

Each sub-sample selected for INAA was freeze-dried at 1 Pa, – 20 °C shelf temperature and – 50 °C condenser temperature, for 5 d. Five of the 15 tissues were subjected to additional drying because the tissues did not appear to be dry after 5 d. Subsequent drying, for an additional 5 d, did not result in significant changes in the values for the ratios of dry to wet mass for these tissues. Determination of total extractable organic material showed that those kidney tissues that did not appear to be dry contained greater amounts of lipid than the other tissues.

The wet appearance of these sub-samples was probably due to a relatively high lipid content rather than to insufficient drying. The ratios of dry mass to wet mass, or conversion factors (C.F.), for the 15 kidney and 15 liver tissue samples together with the animal and tissue identification codes. Two disks, approximately 200 mg each, were formed from each of the dried, powdered subsamples using a commercially available stainless steel die and hydraulic press.

Two disks were also formed from SRM 1566a Oyster Tissue, and one from a pilot whale liver tissue homogenate (Whale Liver Homogenate I) that is used for quality assurance in the analysis of marine mammal tissues. The mass of each disk was recorded and all were packaged in acid-washed linear polyethylene (LPE) film in preparation for irradiation. Disks were also formed from standards consisting of filter papers containing known amounts of each element. Analyses of short-lived products of neutron irradiation were performed sequentially over the course of several days.

Each sample or aliquot of SRM was irradiated, together with one of the standards, for 120 s in the NIST reactor pneumatic tube irradiation facility, RT-4, at reactor power of 20 MW. After the irradiation, each sample, control, and standard was repackaged in clean LPE film for counting. Each sample was placed at a distance of 18 cm from a germanium detector, and gamma rays were collected for 300 s after a decay time of approximately 120 s for determination of Mg, Cl, Ca, V, Cu, Br, and I. Gamma rays were collected again, after 2 to 3 h of decay, for 20 min at a distance of 8 cm for the determination of Na, K, and Mn. Analyses of intermediate-lived and long-lived products of neutron irradiation were performed after the analyses of short-lived products. Samples and standards were packaged individually in bags formed from LPE film and placed together in a polyethylene irradiation vessel or rabbit. For the assay of intermediate- and long-lived nuclides, each rabbit was irradiated in a pneumatic rabbit tube irradiation facility at a reactor power of 20 MW for 16 h.

After a decay time of approximately 6 d, gamma rays from intermediate-lived nuclides (As, Mo, Cd, La, Sm, and Au) were collected for 2 h from samples positioned 15 cm from an intrinsic germanium detector. After decay times of 4 to 8 weeks, gamma rays from long-lived nuclides (Sc, Fe, Co, Zn, Se, Rb, Sr, Ag, Sn, Sb, Cs, Ba, Ce, Eu, Tb, Hf, Ta, Th, and U) were collected for 8 h from samples positioned approximately 5 cm from the same detector. Spectral reduction and data analysis were accomplished using a μVAX 3400 computer with Nuclear Data peak search and activation analysis software. Quantitative evaluation was done by the comparator method using all standards from the individual irradiations. Nuclear Data software was used to calculate "standard constants" (the ratio of the amount of activity of a given nuclide that was present immediately after irradiation to the mass of the element, Ao/g) and these constants were used to determine element concentrations of the samples. Mass fractions of the following elements were at or below the INAA detection limits in these tissues: Sc, V, Sr, Mo, Cd, Sn, Sb, I, Ba, La, Ce, Sm, Eu, Tb, Hf, Ta, Au, Th, and U. INAA detection.

Sources of uncertainty for this analysis included uncertainty of counting statistics, uncertainty in the element content of the standards, uncertainty in dry mass of materials, and any differences between samples and standards in irradiation and counting geometries. Analysis of portions of SRM 1566a and Whale Liver Homogenate (I) were included for the purpose of quality assurance.

ICP-MS Analysis of Cd, Hg, Pb, Sb, Tl, Th, Mn, Sn, Cu, Ag and Zn in Liver and Kidney Samples

Frozen sub-samples of dolphin tissues weighing between 0.3 g and 0.7 g were transferred to tared 100 mL PFA-Teflon containers using a ceramic spatula. The spatula was cleaned with high-purity water, wiped with a metal-free cloth, and chilled before use. The weighing process was performed rapidly to prevent the defrosting and gelling of the tissue samples. Each container of dolphin tissue was sampled at least two times so that duplicate sample preparations could be made. An acid mixture containing 4 mL of HNO_3 and 2 mL of $HClO_4$ was added to each container, which was then sealed. The samples were dissolved by microwave digestion. After digestion, the sample containers were cooled, and the samples transferred to 100-mL low-density-polyethylene bottles and diluted to a final volume of approximately 100 mL with high-purity water containing 1% $K_2Cr_2O_7$. The mass and density of each final solution was used to calculate the analyte concentration. The above elements were determined by inductively coupled plasma mass spectrometry (ICPMS) using a Perkin Elmer in the TotalQuant II mode and external standards. As a quality assurance check on the ICP-MS measurements, Pb was determined on selected subsamples by electrothermal atomic absorption spectrometry (ETAAS) and Hg was determined by cold-vapour atomic absorption spectroscopy (CVAAS).

BIOLOGICAL INFORMATION

Total length (cm) and animal gender are presented in Table. Of the 15 animals, 6 were male and 9 were female; 3 females were pregnant and all fetuses were female. The most common pathological conditions found during the necropsies were ectopic or displaced spleen (9 of 15 animals), pneumonia (8 animals), and fibroid kidneys (7 animals).

Table. Individual Animal and Sample Information

NBSB Number	Age (years)	Length (cm)	Lactating (L) or with fetus (F)	Blubber Lipid (per cent)	Kidney Lipid (per cent)
Male					
RTDL-012	0.5	136	– –	54.9	5.9
RTDL-014	2.5	185	– –	60.0	7.19
RTDL-004	3	199	– –	55.8*	5.92
RTDL-006	4.5	207	– –	73.3	6.13
RTDL-002	10	229	– –	44.4	6.72
RTDL-005	na**	235	– –	42.8	20.5
Female					
RTDL-001	0.8	183	– –	66.3	5.25
RTDL-011	3.5	202	– –	58.1	9.3
RTDL-015	4	177	– –	45.4	6.27
RTDL-010	10	209	– –	54.2	12.2
RTDL-009	7.5	228	– –	61.1	9.6
RTDL-008	22	221	L	49.1	29.8
RTDL-003	25	235	F	38.0	15.8
RTDL-013	42	253	F	50.4	45.2
RTDL-007	na	241	F	41.1	32.4

The mean lipid (total non-volatile solvent extractable material) content of the female roughtoothed dolphin blubber samples was 51.5% with a range from 38.0% to 66.3% and in males was 55.1% with a range from 42.8% to 73.3%. To assess repeatability of lipid determinations, lipid content was measured in triplicate subsamples of RT-004 by NIST Charleston and determined to be 55.8% (1.3%, 1 SD).

The lipid content of RTDL-004 blubber was also determined by the NIST Gaithersburg laboratory and found to contain 56.9% lipid. Lipid content was measured in three samples of SRM 1945 and was determined to be 70.1% (0.1%), compared to the certified value of 74.3% ± 0.4%. The per cent total extractable organic content of the kidney tissues ranged from 5.25% to 45.2% (mean = 14.5%; median 9.27%) and that of the liver tissues ranged from 3.29% to 6.97% (mean = 4.88%; median = 4.87%).

BLUBBER ANALYSES QUALITY CONTROL

Several mechanisms were used to assess the accuracy and precision of data and to evaluate the comparability of data generated by the NIST Charleston and NIST Gaithersburg laboratories. POPs were determined by each laboratory in three aliquots of SRM 1945 that were analysed with the rough-toothed dolphins. Duplicate portions of six rough-toothed dolphin blubber specimens (RTDL-001 through RTDL-006) were analysed by both the NIST Charleston and Gaithersburg laboratories. The largest deviation was observed for PCB 18, where the relative deviation ranged from 33% in RDTL-003 to 164% in RTDL-004. The reason for the deviation was not known as the values for PCB 18 determined in SRM 1945 by both NIST Charleston and NIST Gaithersburg were within the 95% confidence interval of the certified value for this compound (4.48 ng/g wet mass ± 0.88 ng/g wet mass). Repeatability within a batch was assessed by the NIST Charleston Laboratory through three analysis of RTDL-004 made in separate batches. The average per cent relative standard deviation of measurements made on this sample was 6.42% (1.04% to 17.3%).

PCBS AND ORGANOCHLORINE PESTICIDES

Many target compounds including PCBs, DDTs, chlordanes, mirex, dieldrin, HCHs, and HCB were detected in the nine female and six male rough-toothed dolphins analysed. Organochlorine pesticides concentrations in the rough-toothed dolphin samples. All results and comparisons to other studies are given on a lipid-mass basis unless otherwise noted. The results from the NIST Gaithersburg and NIST Charleston analyses were averaged. PCBs were the organochlorine compounds found in greatest concentrations in the rough-toothed dolphins in this study. The sum of PCBs (ΣPCBs; sum of 31 individual congeners or congener groups) in female dolphin samples averaged 25,900 ng/g (range: 1,310 ng/g to 49,400 ng/g). ΣPCBs in males averaged 46,900 ng/g ΣPCBs (range: 24,900 ng/g to 73,600 ng/g). PCB congener 153 contributed most to the ΣPCBs with a mean per cent contribution of 24% (range: 22% to 29%) in females, and 26% in males (range: 24% to 29%). Other PCB congeners detected at relatively high concentrations included PCBs 99, 101, 118, 138, 180, and 187. When compared to cetaceans from studies in other geographic regions, the rough-toothed dolphins from this sample set have lower levels of ΣPCBs. For example, higher concentrations of ΣPCBs have been observed in male Risso's dolphins (Grampus griseus; mean: 128,000 ng/g) inhabiting coastal Japanese waters and in stranded male striped dolphins from the western Mediterranean Sea.

Kuehl and Haebler reported that male bottlenose dolphins that stranded on the Gulf Coast of Texas had a mean ΣPCB concentration of 93,000 ng/g, which is higher than that of the rough-toothed dolphins. Other studies examining stranded bottlenose dolphins from the Texas and the Atlantic coasts reported higher mean ΣPCB concentrations-36,100 ng/g and 60,600 ng/g, respectively. With such a limited data set, and because so little is known about the relationship between the cycling of contaminants and the pelagic food web in the Gulf of Mexico, any conclusions about the differences in mean ΣPCBs between coastal bottlenose dolphins and pelagic rough-toothed dolphins must be made with caution. We speculate that this difference is attributable primarily to habitat differences, as the near-shore waters of the Gulf of Mexico are in much closer proximity to potential point sources of PCBs resulting in higher contamination of the coastal food web and dolphin prey species. The pelagic rough-toothed dolphins' prey may have lower exposures and concentrations of PCBs, accounting for the difference in the ΣPCBs. Additionally, the rough-toothed dolphins feed at a lower tropic level than the bottlenose dolphins, which would result in less biomagnification of contaminant levels.

The suggested toxic threshold concentration of PCBs in marine mammal blubber is 17,000 ng/g lipid mass. Additionally, results from Lahvis *et al*. indicate that even relatively low levels of PCBs and DDTs can have negative effects on immune system function. Based on this criterion, it is likely that for the population represented by this limited sample group, PCBs pose a health risk. Unfortunately, little is understood about the mechanisms of cumulative actions of the complex mixtures of organochlorines to which these dolphins are exposed, and how these compounds impact the animals' health status. Therefore, it is not currently possible to determine whether the health conditions and ultimate deaths of these dolphins were related to body burdens of POPs. The DDT group compounds (2,4'-DDD, 4,4'-DDD, 2,4' DDE, 4,4'-DDE, 2,4'-DDT, and 4,4'-DDT) were the organochlorine pesticide compounds present in the highest concentrations. In females, the average ΣDDT concentration was 14,100 ng/g ranging from 298 ng/g to 34,900 ng/g .

In males, the mean ΣDDT was 22,100 ng/g, ranging from 10,300 ng/g to 37,900 ng/g . 4,4'-DDE was the DDT group compound present in the highest concentration, contributing 71% (range: 48% to 89%) to ΣDDTs in females, and 79% (range: 73% to 92%) in males. The male rough-toothed dolphins had mean ΣDDT concentrations similar to those reported previously in several odonteceti species including Dall's porpoise (Phocoenoides dalli) from both the northeastern and northwestern Pacific (mean: 17,900 ng/g) and Pacific white-sided dolphins (Lagenorhynchus obliquidens) from the North Pacific. Coastal bottlenose dolphins from the Gulf of Mexico had a mean ΣDDTs of 15,300 ng/g, which was lower than the mean level found in the pelagic rough-toothed dolphins.

However, the maximum concentrations of ΣDDTs in bottlenose dolphins from both the Gulf of Mexico (74,600 ng/g) and the Atlantic (80,000 ng/g) were more than double those measured in the rough-toothed dolphins. Prudente *et al.* Salata *et al.* Kuehl *et al.* and Kuehl and Haebler all reported that 4,4'-DDE contributed the most to ΣDDTs in all cetacean species they studied, consistent with our results. To better evaluate exposure to other DDT compounds and currentuse DDT, the percentage ratio of 4,4'-DDT to 4,4'-DDT + 4,4'-DDE was calculated. In females, the mean ratio was 18 (range: 13 to 38) and in males, the mean ratio was 9.2 (range: 3.0 to 16). Interestingly, the ratios were much higher in the rough-toothed dolphins than those from bottlenose dolphins that stranded in the Gulf of Mexico.

The mean ratio in bottlenose dolphins that were analysed by Salata *et al.* was 4.1, while the mean in dolphins analysed by Kuehl and Haebler was 4.2. This difference may be attributable to two reasons. One explanation may be that the range of the rough-toothed dolphins in the Gulf of Mexico includes areas that are more heavily contaminated by current use DDT than the U.S. Gulf Coastal range of bottlenose dolphins. Mexico, which forms the western boundary of the Gulf of Mexico, currently uses DDT for mosquito control in amounts that exceed DDT usage by other Latin American countries. If the rough-toothed dolphins feed in this region, it may explain their elevated levels of parent DDT compounds. A second explanation may be that rough-toothed dolphins have less ability than bottlenose dolphins to metabolize DDT compounds, perhaps because this species lacks the CYP2B gene that transcribes the cytochrome P450 mono-oxygenase isozyme responsible for metabolism of these compounds.

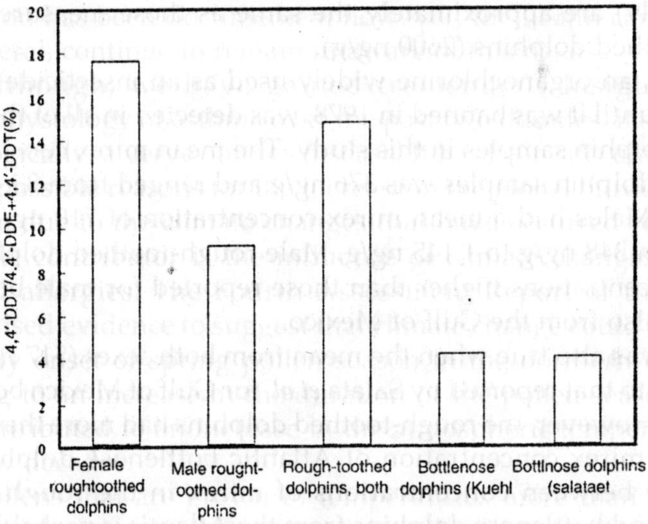

Fig. Comparison of the Ratios of 4,4'-DDT to 4,4'-DDT + 4,4'-DDE (%) in Dolphin Species from the Gulf of Mexico.

Following DDTs, the sum of chlordanes (Σchlordanes; the sum of heptachlor, heptachlo: epoxide, cis-nonachlor, *trans*-nonachlor, *cis*-chlordane, trans-chlordane, and oxychlordane) was the organochlorine pesticide group detected at highest concentration. Ochlordanes averaged 3,670 ng/g in females (40.3 ng/g to 12,400 ng/g) and 4,060 ng/g in males (1,620 ng/g to 8,200 ng/g). *Trans*-nonachlor contributed the most to Σchlordanes, with a mean percentage of total in females of 61%, ranging from 52% to 80%, and in males of 70%, ranging from 56% to 82%. The highest mean concentrations of Σchlordanes found by Prudente *et al.* were in male Risso's dolphins from Japanese coastal waters; mean Σchlordanes were 12,800 ng/g, or approximately three times higher than the mean concentration measured in this set of roughtoothed dolphins.

This difference may be attributable to global fractionation of chlordane compounds resulting in higher levels at higher latitudes, and therefore greater contamination of prey species and ultimately top tier predators, a common trend with many organochlorines. Salata *et al.* reported a mean Σchlordanes concentration of 3,890 ng/g lipid mass in bottlenose dolphins of both sexes from the Gulf of Mexico, which was lower than the measured mean Σchlordanes concentration in the roughtoothed dolphins from this study. However, if only levels of *trans*-nonachlor are compared, the mean concentrations from bottlenose dolphins inhabiting the coastal waters of both the Gulf of Mexico and the Atlantic (2170 ng/g lipid mass and 2930 ng/g lipid mass respectively) are approximately the same as those measured in the rough-toothed dolphins (2600 ng/g).

Mirex, an organochlorine widely used as an insecticide and fire retardant until it was banned in 1978, was detected in all of the rough-toothed dolphin samples in this study. The mean mirex concentration in female dolphin samples was 376 ng/g and ranged from 33.4 ng/g to 912 ng/g. Males had a mean mirex concentration of 710 ng/g with a range from 348 ng/g to 1,145 ng/g. Male rough-toothed dolphins had mirex concentrations higher than those reported for male bottlenose dolphins also from the Gulf of Mexico.

This was also true when the mean from both sexes (517 ng/g) was compared to that reported by Salata *et al.* for Gulf of Mexico bottlenose dolphins. However, the rough-toothed dolphins had more than double the mean mirex concentration of Atlantic bottlenose dolphins. The difference between concentrations of mirex in the rough-toothed dolphins and bottlenose dolphins from the Atlantic is probably due to extensive mirex application in the southeastern U.S. for fire ant control and because of the compound's chemical properties. Mirex is very hydrophobic and non-volatile, so is therefore unlikely, once it has entered the environment, to volatilize and form a latitudinal concentration gradient as is seen with many other POPs.

The difference in mirex concentrations between animals from the Gulf of Mexico and the Atlantic may make mirex potentially useful as a marker compound to discriminate stocks of rough-toothed dolphins in these two regions. However, there are no data available on contaminant levels in rough-toothed dolphins from the Atlantic, so comparisons are not currently possible. Dieldrin was detected in all of the rough-toothed dolphin samples analysed. Dieldrin concentrations in female dolphin samples had a mean of 467 ng/g and ranged from 18.4 ng/g to 1,840 ng/g. Males had a mean dieldrin concentration of 306 ng/g and ranged from 189 ng/g to 487 ng/g. By comparison, adult male bottlenose dolphins analysed by Kuehl and Haebler had a higher mean dieldrin concentration (1,080 ng/g) than that found in the rough-toothed dolphins from this study.

Salata *et al.* and Kuehl *et al.* also reported higher mean dieldrin concentrations in bottlenose dolphins of both sexes from the Gulf of Mexico and Atlantic (547 ng/g and 1450 ng/g respectively). HCHs were detected at relatively low levels in all individuals analysed. In females, the mean concentration of ΣHCHs (α-, β-, and γ-HCH) was 69.5 ng/g, ranging from 5.29 ng/g to 268 ng/g. The ΣHCHs in males averaged 23.0 ng/g and concentrations ranged from 14.5 ng/g to 45.1 ng/g. In the majority of females, β-HCH contributed the most to ΣHCHs, with a mean contribution of 66% to the ΣHCH (range from 38% to 100%). However, in RTDL-003, only γ-HCH was detected, while in RTDL-008, α-HCH contributed 100% to the ΣHCHs.

In males, β-HCH contributed the most to ΣHCHs, contributing an average of 74.5% to the ΣHCH (range from 64% to 85%). The rough-toothed dolphins had lower concentrations of ΣHCHs compared to concentrations found in Dall's porpoise from the Japan Sea. Coastal bottlenose dolphins from the Gulf of Mexico had mean β-HCH concentrations of 64.2 ng/g. This was higher than the mean β-HCH concentration found in the rough-toothed dolphins. HCH patterns in the rough-toothed dolphins, with β-HCH being the dominant congener, were similar to those found by reported by Prudente *et al.* and Salata *et al.* Lower ΣHCH concentrations in rough-toothed dolphins relative to animals from waters adjacent to Asia were not surprising because most current HCH use is in India.

Additionally, Iwata *et al.* determined that there was a latitudinal concentration gradient with HCHs resulting in higher levels at higher latitudes, which would explain the relatively low levels of HCHs detected in these sub-tropical animals. HCB concentrations were among the lowest of all POPs detected. Mean lipid mass HCB concentrations in female dolphins was 48.4 ng/g and ranged from < 1 ng/g to 70.9 ng/g . Males had a mean HCB concentration of 48.2 ng/g and ranged from 20.8 ng/g to 79.5 ng/g. The highest concentration of HCB detected by Prudente *et al.* was in Dall's porpoises (Phocoenoides dalli) from the Sea of Japan, with a mean HCB concentration of 1,430 ng/g . These authors suggest that HCB distribution is heavily dependent upon temperature due to its high volatility, with HCB partitioning mainly to cold areas or higher latitudes. This could explain why animals from the sub-tropical Gulf of Mexico, such as the rough-toothed dolphins, or bottlenose dolphins (mean: 510 ng/g) have lower concentrations of HCB.

The variability of levels of organochlorines caused by gender and age is an interesting feature of this data set. Unfortunately, the data set is small (n = 15) and limited by the absence of some age groups, thus limiting a complete interpretation. For instance, older males, typically having the highest POP concentrations, are not represented. However, age and gender trends in the POP data can still be observed even with this limitation. One interesting feature of the data, and one of concern, is the relatively high POP concentrations in the immature dolphins. It is well established that mothers transfer organochlorines to their offspring via lactation. While the authors could not find any published information on the nursing duration of rough-toothed dolphin calves, there is information on bottlenose dolphins.

Assuming that the botttlenose dolphin calves nurse for similar lengths of time, rough-toothed dolphin calves under the age of 18 months were likely receiving most or all of their nutrients from their mothers' milk. For calves whose mothers have had previous offspring, the risk of negative health effects from POPs is diminished as females may offload up to 80% of their body burden to their first-born calves. This would explain the elevated levels seen in the young rough-toothed dolphins.

The elevated levels in RTDL-007, a pregnant female, suggest that she was carrying her first calf and had not yet off-loaded her body burden to a previous offspring. Unfortunately, none of the neonates were available for analysis, which could have provided information about parturitional transfer of POPs in this species. Previously, it has been suggested that the predominantly lactational transfer of POPs to first-born calves may have serious health implications including increased mortality. Considering that so little is known about many of the details of the rough-toothed dolphins life history, this is an area warranting further study.

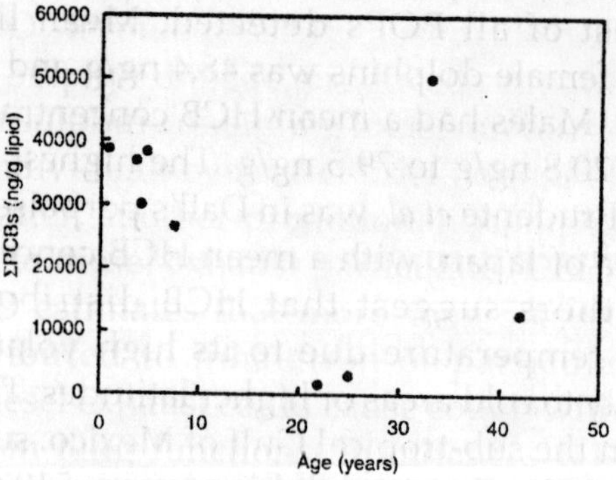

Fig. Age Versus ΣPCB Congeners (ng/g lipid mass) for Female Rough-toothed Dolphins.

LIVER AND KIDNEY ANALYSES

Results of INAA and ICP-MS analyses to determine inorganic constituents in kidney and liver tissues of the 15 rough-toothed dolphins are discussed below. For each method, analyses were performed on duplicate sample portions and results from analyses on these individual portions. Duplicate portions of SRM 1566a Oyster Tissue and of Whale Liver Homogenate I were used as control materials for INAA and one portion of Whale Liver Homogenate I was used for the ICP-MS analyses.

In general, values determined by INAA and ICP-MS agree with certified values, reference values, or literature values. Analysis of Duplicate Subsamples: For most elements, the values from INAA and ICP-MS of duplicate portions agree within the total uncertainties listed for each method. Poor agreement between duplicate portions was observed for Pb in all tissues, Ca in three kidney specimens and two liver specimens, Cd in one liver specimen, and for several elements in the kidney tissue with highest lipid content (RTDL-013). Values for Pb in a given sub-sample varied by as much as two orders of magnitude. It is unlikely that the tissues are inhomogeneous with respect to lead. It is probable that the observed inhomogeneity is the result of contamination of the tissues either during collection or processing of the sub-samples.

Identification of the source of contamination is under investigation. For this reason, the values for Pb are not included in this report. Calcium values from duplicate portions showed poor agreement in several of the specimens. Poor agreement for duplicate portions was observed for liver tissues from RTDL-003 and RTDL-007 and for kidney tissues from RTDL-005, -007, and -008. The kidney tissues of these animals contained greater amounts of lipid material than the others. Although kidney lipid levels were highest for RTDL-013, Ca values in this specimen were below the INAA detection limit of 16 mg/kg. This type of apparent Ca inhomogeneity was found in beluga whale kidney tissues that were known to contain Ca nodules. Tissue mineralization is not uncommon and could result in tissues that contain small Ca deposits that are not homogeneously distributed, even after cryogenic homogenization.

The values from the duplicate sub-samples of the kidney tissue containing the highest amount of lipid, from animal RTDL-013, showed poor agreement for the elements Na, Cl, K, and Br which are short-lived NAA products. For this reason, an additional sub-sample was included in the subsequent analysis of long-lived products of NAA. Values among these three portions were inconsistent. The portions from this tissue were too oily to be easily removed from the LPE bags after the long irradiation. It is possible that inconsistent LPE film element content contributed to the observed variation; however it is more likely that the sub-samples from this tissue are inhomogeneous.

Values for several of the electrolytes in duplicate portions of the tissues with high lipid content were inconsistent and were generally lower than those in the other tissues. Electrolytes are hydrophilic rather than lipophilic so that these results may be expected.

ELECTROLYTES: NA, MG, CA, CL, K, BR

Although most researchers do not determine electrolyte levels in marine mammal tissues, results of this study were compared with other INAA results from analysis of marine mammal tissues from the NBSB. In general, INAA of marine mammal liver tissues has shown that concentrations of electrolytes vary little from animal to animal, or among species. Levels of most of the electrolytes in liver and kidney of rough-toothed dolphins are consistent with this finding.

Values of the relative standard deviation for Na, Mg, Cl, K, and Br for these 15 rough-toothed dolphin liver tissues ranged from 9% to 13%. Little variation was observed because these elements are essential and are regulated biochemically. Levels of Na, Mg, Cl, K, Br, and Cs in these tissues are similar to those found in NBSB tissues of several other odontocetae including white-sided dolphin, pilot whale, harbor porpoise, and beluga whale. Levels of Ca were high and amounts in duplicate portions varied greatly in kidneys of three rough-toothed dolphins and in livers of two (RTDL-003, RTDL- 007). The three kidney specimens were those with high lipid content (20% to 32%).

Calcium values were below the INAA detection limit (of about 16 mg/kg) in the specimen (RTDL-013) containing the largest amount of extractable organic material (45%). Calcium levels found in the rough-toothed dolphins were similar to those found in the beluga whales from Cook Inlet which contained calcified nodules and were inhomogeneous with respect to Ca. Tissue calcification has been associated with connective tissue kidney lesions in humans and with carcinoma.

ESSENTIAL TRACE METALS: MN, CO, CU, ZN, SE

Levels of four of these essential trace elements in liver tissues of rough toothed dolphins also showed little animal to animal variation. Values for the relative standard deviation for Mn, Co, Cu, and Zn ranged from about 10% to 25%. As was the case with electrolytes, this relatively narrow range of values was expected since these elements are essential and are regulated biochemically.

The ranges of values found for these elements in rough-toothed dolphin liver and kidney were generally similar to those reported for other marine mammal species. Although Se is an essential element, much greater animal-to-animal variation was observed for Se concentrations than for the other essential elements. Selenium mass fractions ranged from 2.9 mg/kg to 122 mg/kg in liver tissues and from 2.6 mg/kg to 11.1 mg/kg in kidney tissues. Values for the relative standard deviation of selenium were 40% for concentrations in kidney and 110% for liver. Similar ranges have been reported for other marine mammal species. This comparatively wide range of values was probably due to the role of selenium in metal detoxification.

Selenium or seleno-proteins may assist in the removal of several toxic metals so that Se concentrations often increase with increasing concentrations of other metals. Koemann *et al.* first reported a linear correlation between Se and Hg in liver tissues. Since then, many other researchers have also observed this relationship. Hepatic Se concentrations also increase with those of Ag in several species of odontocetae. Selenium values were slightly higher in the rough-toothed dolphin kidneys when compared with levels found in tissues of other species. The Se concentrations in kidney tissues containing a higher amount of lipid appeared to be slightly higher than in tissues with a lower amount of lipid. Selenium values from animals RTDL-003, -005, -007, -008, and -013, ranged from 6.9 mg/kg to 11.1 mg/kg, whereas the range for the remaining tissues was 2.4 mg/kg to 5.9 mg/kg.

Selenium values in the liver tissues from those dolphins were also higher (with a range of 33 mg/kg to 122 mg/kg) in the animals whose kidneys contained higher lipid amounts (RTDL-003, -005, -007, -008 and -013) compared to a range of 2.9 mg/kg to 28 mg/kg for the remaining animals. The number of tissues analysed was insufficient to determine whether this difference was significant. Similar patterns were observed for Hg, which is discussed below.

POTENTIALLY TOXIC METALS: AS, AG, CD, HG

Levels of As in rough-toothed dolphins were below the INAA detection limit of 0.06 mg/kg in 14 of the 15 kidney tissues analysed and ranged from 0.2 mg/kg to 0.6 mg/kg in rough-toothed dolphin liver tissues. These ranges were consistent with values reported for other species of marine mammal. Hepatic As levels in roughtoothed dolphins were similar to those found in harbor porpoise and the range overlaps ranges reported for several other species of marine mammals. Levels of Ag, Cd, and Hg in rough-toothed dolphin tissues showed much greater animal-toanimal variation than As.

The relative standard deviations of the average values for these elements ranged from about 50% to well over 100%, as values for a given element span several orders of magnitude. The elements Ag, Cd, and Hg are probably not essential, and have no known specific biochemical regulatory pathways in mammals. These metals often accumulate in liver and kidney tissues over time. Results from ICP-MS showed that renal Cd values ranged from 0.05 mg/kg to 3.94 mg/kg. These values are similar to the levels found in the kidneys of other marine mammals from the NBSB and with selected literature values. Levels of Hg in rough-toothed dolphin kidneys also spanned two orders of magnitude, consistent with ranges observed in other species. Hepatic Hg in roughtoothed dolphins RTDL-003, -007 and -013 represented some of the higher values (175 mg/kg, 180 mg/kg, and 235 mg/kg) observed for tissues from the NBSB.

The hepatic Hg values for rough-toothed dolphins were within the range of values reported for other odontocetae such as Mediterranean striped dolphins (Stenella coeruleoalba;1.2 mg/kg to 1544 mg/kg) and were similar to the values for harbor porpoise reported by Law *et al.*, 1992.

RELATIONSHIP BETWEEN LIVER AND KIDNEY

In general, no strong relationships were found between element concentrations in kidney and liver. Concentrations of Se in liver and kidney appeared to be correlated with a linear correlation coefficient of 0.75 ($p<0.002$). No other significant linear correlation relationships were observed between the two tissues.

ELEMENT ACCUMULATION WITH AGE

Liver and kidney may serve as repositories for some trace elements. Data from this study were analysed to determine whether any elements increased with increasing animal age. Results indicated that Se, Hg, Ag, and Sn accumulated in rough-toothed dolphin liver and that Hg and Se accumulated in kidney tissues.

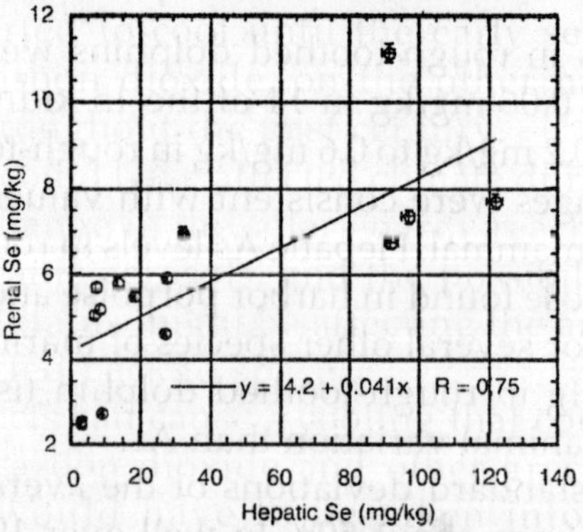

Fig. Renal and Hepatic Se Concentrations (mg/kg) in Rough-toothed Dolphins.

Several researchers observed that levels of Hg and Se were correlated and increase over time in liver tissue of several mammalian species. This relationship was thought to be the result of Se assisting in the detoxification of Hg or its compounds. Results of this work indicated that Se and Hg accumulated with age in both liver and kidney of rough-toothed dolphins.

The linear correlation coefficient (r) for hepatic Se concentrations versus age was 0.77, and for hepatic Hg versus age, $r = 0.96$. The linear correlation coefficient for renal Se concentrations vs. age was 0.95, and that for renal Hg was 0.92. Accumulation of Hg and Se with age was reported in many marine mammal species.

Less is known about levels of Sn in marine mammal tissues. Most researchers do not determine Sn. There are some data on levels of the butyltin compounds in marine mammal tissues. Law *et al.* report that levels of total butyltin compounds in liver tissues of porpoises and grey seals (Halichoerrs grypus) ranged from 0.01 mg/kg to 0.6 mg/kg and levels as high as 10 mg/kg have been reported. Levels of Sn in livers of these animals ranged from 0.08 mg/kg to 1.10 mg/kg. Levels in kidney tissues ranged from the ICP-MS detection limit of 0.02 mg/kg to 0.14 mg/kg. Hepatic Sn was linearly correlated with animal age ($r = 0.95$) indicating accumulation in liver tissues with time. Hepatic Ag concentrations ($r = 0.90$) also increased with animal age. This was also observed for several species of marine mammals from the NBSB.

ELEMENT CORRELATIONS

Data were analysed to determine whether concentrations of any elements were correlated. Positive linear correlations were observed for Na:Cl; Se:Ag; Hg:Se; Ag:Hg; Hg:Sn; and Se:Sn in liver tissue, and for Se:Hg in kidney. Since Se, Ag, Sn, and Hg were found to increase with animal age, it is possible that there were no direct biochemical relationships between these elements, but simply that they all increased with age. The correlation of Na and Cl may reflect electrolyte balance. Positive linear correlation relationships were also observed for Mn:Zn and Rb:K in kidney tissues.

This study has documented persistent organic pollutant (POP) and element concentrations in a pelagic species for which little data of this kind are otherwise available. Levels of the POP compounds discussed above were similar to or lower than those found in other species of odontecetes from the Gulf of Mexico as well as other geographical locations. However, the roughtoothed dolphins had higher percentage ratios of 4,4′-DDT to 4,4′-DDT + 4,4′-DDE than those measured in bottlenose dolphins from the Gulf of Mexico, suggesting higher exposure to currentuse DDTs. Differences in mirex concentrations between rough-toothed dolphins and bottlenose dolphins from the Gulf of Mexico versus bottlenose dolphins from the western Atlantic may prove useful in discriminating between stocks of rough-toothed dolphins from the two regions. Of particular concern were the relatively high concentrations of POPs in the immature dolphins. The contaminant concentrations versus age structure suggests that there is substantial off-loading of females' body burdens to their calves, which is of concern, as little is known about the potential negative effects on individual animals and population levels.

Element contents of liver and kidney tissues are, in general, consistent with levels found in other odontocetae. Levels of Se and Hg in liver tissues were among some of the higher values found in marine mammal tissues from the NBSB, but were similar to those reported for several other species of odontocetae. The elements Se, Ag, Sn, and Hg accumulate in rough-toothed dolphin liver with age. Results also indicate that Se and Hg accumulate in rough-toothed dolphin kidney. These findings may reflect the protective mechanism of Se in detoxification or removal of Hg in these tissues.

Levels of hepatic Hg and Se were higher in the rough-toothed dolphins with Ca inhomogeneity and higher lipid content in the kidney. Additional study is required to determine the significance of this finding. The mortality event of rough-toothed dolphins in 1997 provided a limited sample set from which much can be learned, but there are still many unanswered questions: e.g., What are contaminant levels in older adult males? How do levels and contaminant profiles from Atlantic rough-toothed dolphins compare with those for rough-toothed dolphins from the Gulf of Mexico? Unfortunately, to address these questions, researchers are dependent on mass-mortality events from which to obtain samples.

The MMHSRP administered by NOAA provides the support and coordination for a systematic investigation of mass stranding events and for collection and banking of tissues from these animals so that researchers will have the opportunity to address these questions in the event of future rough-toothed dolphin stranding events.

Chapter 11

Global Warming

Global warming poses one of the most serious threats to the global environment ever faced in human history. Yet by focusing entirely on carbon dioxide emissions, major environmental organizations have failed to account for published data showing that other gases are the main culprits behind the global warming we see today. As a result, they are neglecting what might be the most effective strategy for reducing global warming in our lifetimes: advocating a vegetarian diet.

GLOBAL WARMING AND CARBON DIOXIDE

The environmental community rightly recognizes global warming as one of the gravest threats to the planet. Global temperatures are already higher than they've ever been in at least the past millennium, and the increase is accelerating even faster than scientists had predicted. The expected consequences include coastal flooding, increases in extreme weather, spreading disease, and mass extinctions. Unfortunately, the environmental community has focused its efforts almost exclusively on abating carbon dioxide (CO_2) emissions. Domestic legislative efforts concentrate on raising fuel economy standards, capping CO_2 emissions from power plants, and investing in alternative energy sources.

Recommendations to consumers also focus on CO_2: buy fuel-efficient cars and appliances, and minimize their use. This is a serious miscalculation. Data published by Dr. James Hansen and others show that CO_2 emissions are not the main cause of observed atmospheric warming. Though this may sound like the work of global warming skeptics, it isn't: Hansen is Director of NASA's Goddard Institute for Space Studies who has been called "a grandfather of the global warming theory."

He is a longtime supporter of action against global warming, cited by Al Gore and often quoted by environmental organizations, who has argued against skeptics for subverting the scientific process. His results are generally accepted by global warming experts, including bigwigs like Dr. James McCarthy, co-chair of the International Panel on Climate Change's Working Group II. The focus solely on CO_2 is fueled in part by misconceptions. It's true that human activity produces vastly more CO_2 than all other greenhouse gases put together. However, this does not mean it is responsible for most of the earth's warming. Many other greenhouse gases trap heat far more powerfully than CO_2, some of them tens of thousands of times more powerfully. When taking into account various gases' global warming potential—defined as the amount of actual warming a gas will produce over the next one hundred years—it turns out that gases other than CO_2 make up most of the global warming problem.

Even this overstates the effect of CO_2, because the primary sources of these emissions—cars and power plants—also produce aerosols. Aerosols actually have a cooling effect on global temperatures, and the magnitude of this cooling approximately cancels out the warming effect of CO_2. The surprising result is that sources of CO_2 emissions are having roughly zero effect on global temperatures in the near-term! This result is not widely known in the environmental community, due to a fear that polluting industries will use it to excuse their greenhouse gas emissions.

For example, the Union of Concerned Scientists had the data reviewed by other climate experts, who affirmed Hansen's conclusions. However, the organization also cited climate contrarians' misuse of the data to argue against curbs in CO_2. This contrarian spin cannot be justified. While CO_2 may have little influence in the near-term, reductions remains critical for containing climate change in the long run. Aerosols are short-lived, settling out of the air after a few months, while CO2 continues to heat the atmosphere for decades to centuries. Moreover, we cannot assume that aerosol emissions will keep pace with increases in CO_2 emissions. If we fail start dealing with CO_2 today, it will be too late down the road when the emissions catch up with us.

Nevertheless, the fact remains that sources of non-CO_2 greenhouse gases are responsible for virtually all the global warming we're seeing, and all the global warming we are going to see for the next fifty years. If we wish to curb global warming over the coming half century, we must look at strategies to address non-CO_2 emissions. The strategy with the most impact is vegetarianism.

METHANE AND VEGETARIANISM

By far the most important non-CO_2 greenhouse gas is methane, and the number one source of methane worldwide is animal agriculture. Methane is responsible for nearly as much global warming as all other non-CO_2 greenhouse gases put together. Methane is 21 times more powerful a greenhouse gas than CO_2. While atmospheric concentrations of CO_2 have risen by about 31% since pre-industrial times, methane concentrations have more than doubled. Whereas human sources of CO_2 amount to just 3% of natural emissions, human sources produce one and a half times as much methane as all natural sources. In fact, the effect of our methane emissions may be compounded as methane-induced warming in turn stimulates microbial decay of organic matter in wetlands—the primary natural source of methane.

With methane emissions causing nearly half of the planet's human-induced warming, methane reduction must be a priority. Methane is produced by a number of sources, including coal mining and landfills—but the number one source worldwide is animal agriculture. Animal agriculture produces more than 100 million tons of methane a year. And this source is on the rise: global meat consumption has increased fivefold in the past fifty years, and shows little sign of abating.

About 85% of this methane is produced in the digestive processes of livestock, and while a single cow releases a relatively small amount of methane, the collective effect on the environment of the hundreds of millions of livestock animals worldwide is enormous. An additional 15% of animal agricultural methane emissions are released from the massive "lagoons" used to store untreated farm animal waste, and already a target of environmentalists' for their role as the number one source of water pollution in the U.S. The conclusion is simple: arguably the best way to reduce global warming in our lifetimes is to reduce or eliminate our consumption of animal products. Simply by going vegetarian (or, strictly speaking, vegan), we can eliminate one of the major sources of emissions of methane, the greenhouse gas responsible for almost half of the global warming impacting the planet today.

Advantages of Vegetarianism over CO_2 Reduction

In addition to having the advantage of immediately reducing global warming, a shift away from methane-emitting food sources is much easier than cutting carbon dioxide. First, there is no limit to reductions in this source of greenhouse gas that can be achieved through vegetarian diet. In principle, even 100% reduction could be achieved with little negative impact. In contrast, similar cuts in carbon dioxide are impossible without devastating effects on the economy. Even the most ambitious carbon dioxide reduction strategies fall short of cutting emissions by half. Second, shifts in diet lower greenhouse gas emissions much more quickly than shifts away from the fossil fuel burning technologies that emit carbon dioxide. The turnover rate for most ruminant farm animals is one or two years, so that decreases in meat consumption would result in almost immediate drops in methane emissions. The turnover rate for cars and power plants, on the other hand, can be decades. Even if cheap, zero-emission fuel sources were available today, they would take many years to build and slowly replace the massive infrastructure our economy depends upon today.

Similarly, unlike carbon dioxide which can remain in the air for more than a century, methane cycles out of the atmosphere in just eight years, so that lower methane emissions quickly translate to cooling of the earth. Third, efforts to cut carbon dioxide involve fighting powerful and wealthy business interests like the auto and oil industries. Environmental groups have been lobbying for years to make fuel-efficient SUVs available or phase out power plants that don't meet modern environmental standards without success. At the same time, vegetarian foods are readily available, and cuts in agricultural methane emissions are achievable at every meal. Also, polls show that concern about global warming is widespread, and environmental activists often feel helpless to do anything about it. Unless they happen to be buying a car or major appliance, most people wanting to make a difference are given little to do aside from writing their legislators and turning off their lights. Reducing or eliminating meat consumption is something concerned citizens can do every day to help the planet.

Finally, it is worth noting that reductions in this source of greenhouse gas have many beneficial side effects for the environment. Less methane results in less tropospheric ozone, a pollutant damaging to human health and agriculture. Moreover, the same factory farms responsible for these methane emissions also use up most of the country's water supply, and denude most of its wilderness for rangeland and growing feed. Creating rangeland to feed western nations' growing appetite for meat has been a major source of deforestation and desertification in third world countries. Factory farm waste lagoons are a leading source of water pollution in the U.S. Indeed, because of animal agriculture's high demand for fossil fuels, the average American diet is far more CO_2-polluting than a plant-based one.

PREDICTION OF CLIMATE CHANGE

The prediction of climate change due to human activities began with a prediction made by the Swedish chemist, Svante Arrhenius, in 1896. Arrhenius took note of the industrial revolution then getting underway and realized that the amount of carbon dioxide being released into the atmosphere was increasing. Moreover, he believed carbon dioxide concentrations would continue to increase as the world's consumption of fossil fuels, particularly coal, increased ever more rapidly. His understanding of the role of carbon dioxide in heating Earth, even at that early date, led him to predict that if atmospheric carbon dioxide doubled, Earth would become several degrees warmer. However, little attention was paid to what must have been seen to be a rather far-out prediction that had no apparent consequence for people living at that time. Arrhenius was referring to a potential modification of what we now call the greenhouse effect. A simplified explanation of this is as follows.

Shortwave solar radiation can pass through the clear atmosphere relatively unimpeded, but longwave infrared radiation emitted by the warm surface of the Earth is absorbed partially and then reemitted by a number of trace gases—particularly water vapour and carbon dioxide—in the cooler atmosphere above. Because, on average, the outgoing infrared radiation balances the incoming solar radiation, both the atmosphere and the surface will be warmer than they would be without the greenhouse gases. One should distinguish between the "natural" and a possible "enhanced" greenhouse effect.

The natural greenhouse effect causes the mean temperature of the Earth's surface to be about 33°C warmer than it would be if natural greenhouse gases were not present. This is fortunate, for the natural greenhouse effect creates a climate in which life can thrive and humankind can live under relatively benign conditions. Otherwise, the Earth would be a very frigid and inhospitable place. On the other hand, an enhanced greenhouse effect refers to the possible raising of the mean temperature of the Earth's surface above that occurring due to the natural greenhouse effect because of an increase in the concentrations of greenhouse gases due to human activities.

Such a global warming would probably bring other, sometimes deleterious, changes in climate; for example, changes in precipitation, storm patterns, and the level of the oceans. The word "enhanced" is usually omitted, but it should not be forgotten in discussions of the greenhouse effect. Nearly 100 years after the Arrhenius prediction, we are now aware that carbon dioxide in the atmosphere is increasing, with the possibility that it will double by the middle of the next century from the levels at the time of Arrhenius. Post-World War II industrialization has caused a dramatic jump in the amount of carbon dioxide in the atmosphere. As the prospect of considerable change in the atmosphere becomes more real and threatening, new computer models are being applied to the problem. These models take into account the natural processes that must be part of the whole picture to understand what could happen to Earth's climate as carbon dioxide increases. An important aspect of the newer models is their treatment of the "amplifier" or feedback effect, in which further changes in the atmosphere occur in response to the warming initiated by the change in carbon dioxide. In addition to moisture and cloud processes, the newer models are beginning to take into account the role of vegetation, forests, grasslands, and crops in controlling the amount of carbon dioxide that actually will be in the atmosphere.

Along with their role as "sinks" for carbon dioxide, the various types of vegetation in the biosphere have further effects on climate. Plants heat or cool the air around them (through the reflection and absorption of solar radiation and the evaporation process), remove momentum from surface winds, and take up and release moisture into the air (thus contributing to alterations in the hydrologic cycle). In turn, changes in climate will affect the patterns of vegetation growth. For instance, forest stands that require relatively cool conditions may not be able to adjust to the relatively rapid warming that is being predicted for the interiors of continents.

With slow warming, scientists expect that the northern edges of North American forests would creep slowly forward to more-favourable conditions, while the southern edges would give way to grasslands that are better suited to the warmer conditions. With overly rapid warming rates, however, the loss at the southern edge would be more extreme, and the migration at the northern edges would not be able to make up for the loss at the southern edge.

Other feedback effects at work also must be considered. In normal conditions, plant leaves take in carbon dioxide from the air and release moisture to the air as part of the photosynthesis process. The release of moisture through evapotranspiration causes the air to cool. With increasing atmospheric carbon dioxide, one can expect to see a change in plant carbon exchange rates and water relations. This may result in reduced evaporation rates, thus amplifying the summer continental warming. Without plants, the ground and air would become warmer, exacerbating the problem.

GREENHOUSE GASES

To predict climate change, one must model the climate. One test of the validity of predictions is the ability of the climate models to reproduce the climate as we see it today. Elements of the models such as the physics and chemistry of the processes that we know—or think we know—are essential to represent in the models. Therefore, the models have to embody the characteristics of the land and the oceans that serve as boundaries of the atmosphere represented in the models. Models also have to take into account the radiative characteristics of the gases that make up the atmosphere, including the key radiative gas, water vapour, that is so variable throughout the atmosphere. Global records of surface temperature over the last 100 years show a rise in global temperatures (about 0.5°?C overall), but the rise is marked by periods when the temperature has dropped as well. If the models cannot explain these marked variations from the trend, then we cannot be completely certain that we can believe in their predictions of changes to come.

For example, in the early 1970's, because temperatures had been decreasing for about 25 to 30 years, people began predicting the approach of an ice age! For the last 15 to 20 years, we have been seeing a fairly steady rise in temperatures, giving some assurance that we are now in a global warming phase. The major gases in the atmosphere, nitrogen and oxygen, are transparent to both the radiation incoming from the sun and the radiation outgoing from the Earth, so they have little or no effect on the greenhouse warming. The gases that are not transparent are water vapour, ozone, carbon dioxide, methane, nitrous oxide, and the chlorofluorocarbons (CFCs).

These are the greenhouse gases. There has been about a 25% increase in carbon dioxide in the atmosphere from 270 or 280 parts per million 250 years ago, to approximately 360 parts per million today. The record of carbon dioxide in the atmosphere shows a variation as seasons change. This variation is more pronounced in the northern hemisphere, with its greater land area, than in the southern hemisphere because of interactions in the atmosphere caused by vegetation. In the growing season, during daylight, vegetation takes in carbon dioxide; at night and in the senescent season, vegetation releases carbon dioxide. The effect is more pronounced in the northern hemisphere because most of the land on Earth is located there.

MODELING

To understand and predict climate change, the following types of models are needed:

- Socio-economic models that predict future fossil fuel consumption and utilization of alternative fuels. These models depend upon technology, e.g., industrial production methods, energy efficiency, new materials, etc.; public policy and social attitudes, e.g., concern for the environment; and economic development, standard of living and reliance on energy and chemical-based products.

- Chemical-physical-biophysical models of the Earth System that tell us what happens to gases released into the atmosphere, e.g., how much carbon dioxide is taken up by the oceans and the biosphere, and how industrial and agricultural uses of chemicals and natural processes on Earth's surface affect the release of methane, nitrogen oxides, and other greenhouse gases into the atmosphere.
- Coupled ocean-atmosphere models that tell us how the climate system, e.g., temperatures, humidity, clouds, and rainfall, responds to changes in the chemical composition of the atmosphere.

Getting reliable predictions from models is difficult because many of the secondary processes are not understood. For example, when temperatures start to warm because of the direct radiative effect of increasing carbon dioxide, will clouds increase or decrease? Will they let in less radiation from the sun, or more? These secondary processes are important. The direct radiative effect of doubling carbon dioxide is relatively small, and there is not much disagreement on this point among models. Where models conflict is in regard to the secondary, or feedback effects.

Models that predict a very large warming from carbon dioxide show cloud cover changes that greatly amplify the warming effects, while models that predict moremodest warming show that clouds have a small or even damping effect on the warming.

Can we match the observation of temperature trends with the model predictions? The temperature record of the past hundred years does show a warming trend, by approximately 0.5°C. However, the observed warming trend is not entirely consistent with the carbon dioxide change. Most of the temperature increase occurred before 1940, after which Earth started to cool until the early seventies, when warming resumed. Carbon dioxide, on the other hand, has been increasing steadily throughout the past century.

Other factors that could have affected climate during this period include the possible change in the solar energy reaching Earth, the cooling effects of volcanic aerosols, and the possibility that sulfur dioxide and other pollutants might be affecting the amount of solar radiation that is reflected back to space.

Some of these effects can cause a cooling that could counteract the warming due to carbon dioxide and other greenhouse gases. All of these effects would have to be taken into account and appropriately modeled in order to predict the changes that one might expect in the next century.

NASA INVESTIGATIONS OF THE GREENHOUSE EFFECT

Over the past 30 years, a number of satellite missions have been launched to obtain the data about Earth's radiation budget that are critical to understanding the greenhouse effect. Some of these missions are listed in the accompanying table. Another very important aspect of greenhouse investigations has been the development of models.

A number of climate models have been developed by NASA, and one of the most detailed is a General Circulation Model (GCM) developed by the Goddard Institute for Space Studies (GISS) in New York City. A GCM uses extremely high-speed computers to solve the basic equations governing atmospheric motions and processes by numerical techniques. The GISS group, using its model, predicted that the annual global temperature would reach a new record high sometime during the first three years of the 1990's. Indeed, that record was reached in 1990.

However, in June 1991, the Mount Pinatubo volcano erupted and sent 25 to 30 million tons of sulfur dioxide into the stratosphere. There, the sulfur dioxide reacted with water vapour to produce a long-lasting haze of sulfuric acid droplets. The GISS group then inserted the new information into the model, estimated how much sunlight the Pinatubo aerosol cloud would block, and predicted that the global temperature would drop about 0.3°C. Again, the predicted change actually occurred.

These successful climate predictions have been encouraging, and scientists are continuing their research to further increase the credibility of their model predictions. An important need in the further development and verification of climate models is the acquisition, assembly, and analysis of reliable climate data. The highlyaccurate, self-consistent, and long-term data sets that will be acquired by the Earth Observing System (EOS), as part of NASA's Mission to Planet Earth, with a series of satellite launches beginning in 1998, are designed to fulfill that need.

EARTH SCIENCE ENTERPRISE

Table. Selected Missions Leading to an Improved Understanding of the Greenhouse Effect

Mission	Launch	Scienfific Objective
Explorer-7	1959	First satellite radation budget experiment (spinning black and white hemispheres)
Nimbus-2, -3	1996, 1969	Global radiation budget measurements from the Medium Resolution Infrared Radiometer
Earth Radiation		Coordinated radiation budget measurements

Budget Experiment (ERBE)/Earth Radiation Budget Satellite (ERBS)	1984	from the Earth Radiation Budget Experiment sensors in three different orbits
NOAA-9	1984	
NOAA-10	1986	
Tropical Rainfall Measuring Mission	1997	Radiation budget data from the Clouds and Earth's Radiant Energy System (CERES) experiments. This mission will complement CERES experiments on EOS satellites
Earth Observing System (EOS)	1998-	Radiation budget data from CERES to complement experiments on TRMM. A broad spectrum ofphysical and chemical measurements of atmosphere, ocean, and land characteristics from different EOS missions

GLOBAL WARMING, CLIMATE CHANGE, AIR POLLUTION

A delicate balance exists between man and his environment. The impact of changes in the surrounding atmosphere on human health was first reported by Plinius, the Younger, in 73 AD. In his documentation of fatal respiratory consequences of natural air pollution caused by the eruption of Mount Vesuvius in Pompeii, he described succinctly, the effect that the clouds of airborne matter in the atmosphere had on Plinius, the Elder, a scientist as well as the commander of the Roman fleet.

Hypothesising that the dense fog was obstructing the airways, he suggested that this effect was caused by "inflammation of airways". The race for development, coupled with the greed of man, has led to pollution of the environment in several ways, threatening the precarious equilibrium between all objects on the planet, living and non-living.

Industrial revolution led to a surge in the consumption of nature's bounties including hydrocarbonaceous fuels, leading to increase in the emission of carbon dioxide (CO_2) and other noxious gases, never before documented in the evolutionary history of earth. These, along with particulate matter, increase the atmospheric trapping of radiated heat from the sun which is known as the "green-house effect". The resultant abnormal rise in the earth's surface temperature is "Global Warming," and the climatic changes, thus, brought about are termed "anthropogenic climate change". The global rise in the concentration of CO_2 is conspicuously linked to "deforestation" a term coined for the rampant destruction of our forests along with, consumption of natural resources by an exploding population and its insatiable appetite for fossil fuel. For centuries, until the commencement of the industrial revolution in the second half of the 18th Century, the level of atmospheric CO_2 concentration was estimated to have remained stagnant at about 280 ± 10 ppm.

The Mauna Loa Observatory in Hawaii recorded a stupendous increase in CO_2 from 316 ppm in 1959 to approximately 380 ppm in March 2008.3 More importantly, two-thirds of this startling increase in the atmospheric CO_2 concentration has occurred over the last 50 years. The World Metrological Organisation and United Nations Environment Programme (UNEP), in an effort to combat the worsening situation, set up the Intergovernmental Panel on Climate Change (IPCC) in 1988.

In recognition of the strong body of evidence that this panel has painstakingly collated, it was honoured with the Nobel Peace Prize last year. The panel recently released their Fourth Assessment Report which categorically states that the "warming of the climate system is unequivocal, as is now evident from observation of increases in global average air and ocean temperature, widespread melting of snow and ice and rising global average sea level".

With 11 of the last 12 years (1995-2006) being the warmest years since the commencement of the temperature records, the average global surface temperature would increase by 1.1 °C to 6.4 °C causing alteration of planetary ecosystem dynamics. The observational evidence garnered by the Fourth Assessment Report has just been confirmed by a meta-analysis published in mid May by Rosenzweig and colleagues. This meta-analysis was the first to formally link the observed alterations in the physical and biological systems in the continents to the anthropogenic climate changes that have occurred since the 1970s.

The Fourth Assessment Report has already identified three areas in which human health has already been affected by climate change. These are:

- Alteration of distribution of some infectious disease vectors,
- Seasonal distribution of some allergenic pollen species, a
- Increased heat wave related deaths.

That climate change impacts health in many ways was highlighted by the World Health Organization (WHO) when it chose to mark World Health Day on April 7 this year with the theme "Protecting health from climate change". The WHO has estimated that climate change is responsible for around 150,000 deaths annually and 5 million Disability Adjusted Life Years (DALY), mainly in developing countries. With this background, this editorial aims to focus on the possible effects of climate change on allergic asthma. The major environmental pollutants that have been linked to asthma are ozone (O_3), nitrogen dioxide (NO_2), sulphur dioxide (SO_2) and particulate matter (PM) smaller than 2.5 μm in diameter.

The Fourth Assessment Report of the IPCC clearly states that "exposure to elevated concentrations of ozone is associated with increased hospital admissions for pneumonia, chronic obstructive pulmonary disease, asthma, allergic rhinitis and other respiratory diseases, and with premature mortality". Vehicular emissions are the major contributors to the rise in the levels of these pollutants. Incidence of respiratory symptoms has been shown to increase in populations living in areas of high road traffic.

Due to inversion effect, ozone is trapped close to the surface of the earth which when inhaled causes airway inflammation and increases responsiveness to allergens in subjects with asthma and/or rhinitis. It not only increases asthma exacerbations, but has also been associated with higher incidence of asthma in children. This is especially in those children who undergo heavy physical exercise while living in areas of increased concentrations of O_3.

The combustion of fossil-derived fuels, along with vehicular emissions, is an important source of NO_2. Apart from its role in production of O_3, it could possibly cause acute decline in lung functions in asthmatics. Sulphur dioxide, produced primarily from the industrial combustion of sulphur containing fuels, can cause bronchoconstriction in asthmatic individuals even after brief periods of exposure. Air borne particulate matter, another predominant component of air pollution, is largely due to diesel exhaust particulate (DEP).

The WHO estimates that more than 500,000 deaths every year could be attributed to inhalation of particulate matter. Chronic exposure to diesel exhaust could cause cough with sputum production and decrease in lung functions. It enhances IgE synthesis in atopic individuals, increasing the sensitivity of the airways to airborne allergens. The carbonaceous core of the DEPs can adsorb several large molecules and allergens, including pollens, thus prolonging their retention. The change in the morphology of the pollen, brought about by its attachment to the pollutant, enhances its allergenic potential.

There is evidence to suggest that air pollutants affect the allergenicity of air-borne allergens which could possibly lead to increased airway hypersensitivity. The DEPs can also serve as a vehicle for airborne pollens and, thus, facilitating its deposition in the lower airways. This interaction between air pollutants and allergens could lead to an increased allergic inflammatory response in the airways causing a possible upsurge in development of atopy and respiratory allergy. The link between climate changes, aeroallergens and air pollution has been documented; however, the health care providers, in general, continue to remain unaware of this nexus.

Although, we have gained tremendous insights into the pathophysiology of asthma over the past two decades, it still remains, aetiologically, a very complex disease. It has been speculated that one of the possible reasons for the increase in the global burden of asthma may be due to the climate changes that have occurred recently. One possible contributor is the influence of climate changes on pollen-related allergies. The Fourth Assessment Report of the IPCC has recognised evidence to suggest that climate change could have caused an early onset of spring pollen season in the northern hemisphere, leading to an increase in the duration of the pollen season. This has been attributed to an increase in the atmospheric temperature in the last few decades.

Pollens that are grown in temperatures more than usual could possibly have stronger antigenicity. Production of pollen too is enhanced by an increase in atmospheric CO_2 concentration. Fungal spores are also known for exacerbation of allergy and asthma. Currently, there is insufficient evidence to link the effect of climate change to growth of fungi, but it appears likely that fungal biomass may proliferate with rise in both CO_2 and temperature. This would further contribute to the increase in the burden as well as in the severity of asthma.

Moreover, the warming of earth's environment reduces winter months. The cold weather which suppresses fungal sporulation is curtailed. In addition, the lower pollen counts due to reduced flowering during the colder months occur for a shorter period. Environmental scientists have further suggested that the rapidly changing climate could have altered the interaction between air pollutants and respiratory allergens. This phenomenon could be due to the atmospheric air pollutants adhering to the surface of microaerosol suspensions of pollens and other plant allergens which may be present even before the usual peak pollen months.

They then cause morphological change in these antigen carrying agents, thereby rendering them more allergenic. Furthermore, these ultra small particles can travel easily to the lower airways, thus increasing the morbidity in pollensensitive asthmatics. Till date, these aspects were never considered of consequence to human health. Evidence regarding the association of asthma with global warming, climate change and air pollution continues to accumulate at a startling speed. Even in the absence of large scientific trials, several contributory factors, as outlined above, have clearly been implicated.

The pulmonologists must not only be aware of the menacing situation but also must play a proactive role to diminish the consequences of these changes on respiratory health. Since the Vedic Era "Prakriti" and "Brahmand" have dominated the thought processes in India but it appears that we have forgotten our legacy. The environmental effect of global warming on India has already been documented but the health hazards, especially the effect on vector-borne diseases like malaria and dengue, are yet to be considered by the authorities concerned.

The atmosphere that surrounds us is changing at an alarming pace, and this fact is now being increasingly recognised. Even if the concentrations of greenhouse gases were to be miraculously frozen at the current levels, it is estimated that the weather patterns would continue to change, for the worse, for the next hundred years before stabilising. Clearly, this phenomenon is here to stay for our lifetime, and it would be prudent for health care providers in India to be aware of the looming disastrous situation. We should work closely with the experts on global warming in order to minimise the impact on human health. There is an urgent need by the society as well as the leaders to acknowledge the ongoing changes so that steps may be initiated to curb the detrimental effects of the global warming.

THE CARBON DIOXIDE GREENHOUSE EFFECT

In the 19th century, scientists realized that gases in the atmosphere cause a "greenhouse effect" which affects the planet's temperature. These scientists were interested chiefly in the possibility that a lower level of carbon dioxide gas might explain the ice ages of the distant past. At the turn of the century, Svante Arrhenius calculated that emissions from human industry might someday bring a global warming. Other scientists dismissed his idea as faulty. In 1938, G.S. Callendar argued that the level of carbon dioxide was climbing and raising global temperature, but most scientists found his arguments implausible.

It was almost by chance that a few researchers in the 1950s discovered that global warming truly was possible. In the early 1960s, C.D. Keeling measured the level of carbon dioxide in the atmosphere: it was rising fast. Researchers began to take an interest, struggling to understand how the level of carbon dioxide had changed in the past, and how the level was influenced by chemical and biological forces. They found that the gas plays a crucial role in climate change, so that the rising level could gravely affect our future.

Like many Victorian natural philosophers, John Tyndall was fascinated by a great variety of questions. While he was preparing an important treatise on "Heat as a Mode of Motion" he took time to consider geology. Tyndall had hands-on knowledge of the subject, for he was an ardent Alpinist (in 1861 he made the first ascent of the Weisshorn). Familiar with glaciers, he had been convinced by the evidence—hotly debated among scientists of his day—that tens of thousands of years ago, colossal layers of ice had covered all of northern Europe. How could climate possibly change so radically?

One possible answer was a change in the composition of the Earth's atmosphere. Beginning with work by Joseph Fourier in the 1820s, scientists had understood that gases in the atmosphere might trap the heat received from the Sun. As Fourier put it, energy in the form of visible light from the Sun easily penetrates the atmosphere to reach the surface and heat it up, but heat cannot so easily escape back into space. For the air absorbs invisible heat rays ("infrared radiation") rising from the surface. The warmed air radiates some of the energy back down to the surface, helping it stay warm. This was the effect that would later be called, by an inaccurate analogy, the "greenhouse effect." The equations and data available to 19th-century scientists were far too poor to allow an accurate calculation. Yet the physics was straightforward enough to show that a bare, airless rock at the Earth's distance from the Sun should be far colder than the Earth actually is.

Tyndall set out to find whether there was in fact any gas in the atmosphere that could trap heat rays. In 1859, his careful laboratory work identified several gases that did just that. The most important was simple water vapour (H_2O). Also effective was carbon dioxide (CO_2), although in the atmosphere the gas is only a few parts in ten thousand. Just as a sheet of paper will block more light than an entire pool of clear water, so the trace of CO2 altered the balance of heat radiation through the entire atmosphere.

GREENHOUSE SPECULATIONS

The next major scientist to consider the question was another man with broad interests, Svante Arrhenius in Stockholm. He too was attracted by the great riddle of the prehistoric ice ages. In 1896 Arrhenius completed a laborious numerical computation which suggested that cutting the amount of CO_2 in the atmosphere by half could lower the temperature in Europe some 4-5°C (roughly 7-9°F) — that is, to an ice age level. But this idea could only answer the riddle of the ice ages if such large changes in atmospheric composition really were possible.

For that question Arrhenius turned to a colleague, Arvid Högbom. It happened that Högbom had compiled estimates for how carbon dioxide cycles through natural geochemical processes, including emission from volcanoes, uptake by the oceans, and so forth. Along the way he had come up with a strange, almost incredible new idea.

It had occurred to Högbom to calculate the amounts of CO_2 emitted by factories and other industrial sources. Surprisingly, he found that human activities were adding CO_2 to the atmosphere at a rate roughly comparable to the natural geochemical processes that emitted or absorbed the gas. The added gas was not much compared with the volume of CO_2 already in the atmosphere—the CO_2 released from the burning of coal in the year 1896 would raise the level by scarcely a thousandth part.

But the additions might matter if they continued long enough. (By recent calculations, the total amount of carbon laid up in coal and other fossil deposits that humanity can readily get at and burn is some ten times greater than the total amount in the atmosphere.) So the next CO_2 change might not be a cooling decrease, but an increase. Arrhenius made a calculation for doubling the CO_2 in the atmosphere, and estimated it would raise the Earth's temperature some 5-6°C.

Arrhenius did not see that as a problem. He figured that if industry continued to burn fuel at the current (1896) rate, it would take perhaps three thousand years for the CO_2 level to rise so high. Högbom doubted it would ever rise that much. One thing holding back the rise was the oceans. According to a simple calculation, sea water would absorb 5/6ths of any additional gas. (That is roughly true over a long run of many thousand years, but Högbom and Arrhenius did not realise that if the gas were emitted more rapidly than they expected, the ocean absorption could lag behind.) Anyway temperatures a few degrees higher hardly sounded like a bad idea in chilly Sweden. Another highly respected scientist, Walter Nernst, even fantasized about setting fire to useless coal seams in order to release enough CO_2 to deliberately warm the Earth's climate.

Arrhenius brought up about the possibility of future warming in an impressive scientific article and a widely read book. By the time the book was published, 1908, the rate of coal burning was already much higher than in 1896, and Arrhenius suggested warming might appear wihin a few centuries rather than millenia. Yet here as in his first article, the possibility of warming in some distant future was far from his main point. He mentioned it only in passing, during a detailed discussion of what really interested scientists of his time—the cause of the ice ages. Arrhenius had not quite discovered global warming, but only a curious theoretical concept.

An American geologist, T. C. Chamberlin, and a few others took an interest in CO_2. How, they wondered, is the gas stored and released as it cycles through the Earth's reservoirs of sea water and minerals, and also through living matter like forests? Chamberlin was emphatic that the level of CO_2 in the atmosphere did not necessarily stay the same over the long term.

But these scientists too were pursuing the ice ages and other, yet more ancient climate changes—gradual shifts over millions of years. Very different climates, like the balmy age of dinosaurs a hundred million years ago, puzzled geologists but seemed to have nothing to do with changes on a human time scale. Nobody took much interest in the hypothetical future warming caused by human industry.

Experts could dismiss the hypothesis because they found Arrhenius's calculation implausible on many grounds. In the first place, he had grossly oversimplified the climate system. Among other things, he had failed to consider how cloudiness might change if the Earth got a little warmer and more humid. A still weightier objection came from a simple laboratory measurement. A few years after Arrhenius published his hypothesis, another scientist in Sweden, Knut Ångström, asked an assistant to measure the passage of infrared radiation through a tube filled with carbon dioxide.

The assistant ("Herr J. Koch," otherwise unrecorded in history) put in rather less of the gas in total than would be found in a column of air reaching to the top of the atmosphere. The assistant reported that the amount of radiation that got through the tube scarcely changed when he cut the quantity of gas back by a third. Apparently it took only a trace of the gas to "saturate" the absorption — that is, in the bands of the spectrum where CO2 blocked radiation, it did it so thoroughly that more gas could make little difference.

Still more persuasive was the fact that water vapour, which is far more abundant in the air than carbon dioxide, also intercepts infrared radiation. In the crude spectrographs of the time, the smeared-out bands of the two gases entirely overlapped one another. More CO_2 could not affect radiation in bands of the spectrum that water vapour, as well as CO_2 itself, were already blocking entirely. These measurements and arguments had fatal flaws. Herr Koch had reported to Ångström that the absorption had not been reduced by more than

0.4% when he lowered the pressure, but a modern calculation shows that the absorption would have decreased about 1% — like many a researcher, the assistant was over confident about his degree of precision.

But even if he had seen the1% shift, Ångström would have thought this an insignificant perturbation. He failed to understand that the logic of the experiment was altogether false. The greenhouse effect will in fact operate even if the absorption of radiation were totally saturated in the lower atmosphere. The planet's temperature is regulated by the thin upper layers where radiation does escape easily into space. Adding more greenhouse gas there will change the balance. Moreover, even a 1% change in that delicate balance would make a serious difference in the planet's surface temperature. The logic is rather simple once it is grasped, but it takes a new way of looking at the atmosphere—not as a single slab, like the gas in Koch's tube (or the glass over a greenhouse), but as a set of interacting layers.

The subtle difference did not occur to anyone for many decades, if only because hardly anyone thought the greenhouse effect was worth their attention. For after Ångström published his conclusions in 1900, the few scientists who had taken an interest in the matter concluded that Arrhenius's hypothesis had been proven wrong. Theoretical work on the question stagnated for decades, and so did measurement of the level of CO_2 in the atmosphere.

A few scientists dissented from the view that changes of CO_2 could have no effect. An American physicist, E.O. Hulburt, pointed out in 1931 that investigators had been mainly interested in pinning down the intricate structure of the absorption bands (which offered fascinating insights into the new theory of quantum mechanics) "and not in getting accurate absorption coefficients." Hulburt's own calculations supported Arrhenius's estimate that doubling or halving CO_2 would bring something like a 4°C rise or fall of surface temperature, and thus "the carbon dioxide theory of the ice ages... is a possible theory."

Hardly anyone noticed this paper. Hulburt was an obscure worker at the U.S. Naval Research Laboratory, and he published in a journal, the Physical Review, that few meteorologists read. Their general consensus was the one stated in such authoritative works as the American Meteorological Society's 1951 Compendium of Meteorology: the idea that adding CO_2 would change the climate "was never widely accepted and was abandoned when it was found that all the long-wave radiation [that would be] absorbed by CO_2 is [already] absorbed by water vapour." Even if people had recognized this was untrue, there were other well-known reasons to deny any greenhouse effect in the foreseeable future. These reasons reflected a nearly universal conviction that the Earth automatically regulated itself in a "balance of nature." Getting to specifics, scientists repeated the plausible argument that the oceans would absorb any excess gases that came into the atmosphere. Fifty times more carbon is dissolved in sea water than in the wispy atmosphere. Thus the oceans would determine the equilibrium concentration of CO_2, and it would not easily stray from the present numbers.

If somehow the oceans failed to stabilize the system, organic matter was another good candidate for providing what one scientist called "homeostatic regulation." The amount of carbon in the atmosphere is only a small fraction of what is bound up not only in the oceans but also in trees, peat bogs, and so forth. Just as sea water would absorb more gas if the concentration increased, so would plants grow more lushly in air that was "fertilized" with extra carbon dioxide. Rough calculations seemed to confirm the comfortable belief that biological systems would stabilize the atmosphere by absorbing any surplus.

One way or another, then, whatever gases humanity added to the atmosphere would be absorbed—if not at once, then within a century or so—and the equilibrium would automatically restore itself. As one respected expert put it baldly in 1948, "The self-regulating mechanisms of the carbon cycle can cope with the present influx of carbon of fossil origin."

Yet the theory that atmospheric CO_2 variations could change the climate was never altogether forgotten. An idea so simple on the face of it, an idea advanced (however briefly) by outstanding figures like Arrhenius and Chamberlin, had to be mentioned in textbooks and review articles if only to refute it. Arrhenius's outmoded hypothesis persisted in a ghostly afterlife. It found a lone advocate. Around 1938 an English engineer, Guy Stewart Callendar, took up the old idea. An expert on steam technology, Callendar apparently took up meteorology as a hobby to fill his spare time.

Many people, looking at weather stories from the past, had been saying that a warming trend was underway. When Callendar compiled measurements of temperatures from the 19th century on, he found they were right. He went on to dig up and evaluate old measurements of atmospheric CO_2 concentrations. He concluded that over the past hundred years the concentration of the gas had increased by about 10%. This rise, Callendar asserted , could explain the observed warming. For he understood (perhaps from Hulburt's calculation) that even if the CO_2 in the atmosphere did already absorb all the heat radiation passing through, adding more gas would change the height in the atmosphere where the absorption took place. That, he calculated, would make for warming.

As for the future, Callendar estimated, on flimsy grounds, that a doubling of CO_2 could gradually bring a 2°C rise in future centuries. He hinted that it might even trigger a shift to a self-sustaining warmer climate (which did not strike him as a bad prospect). But future warming was a side issue for Callendar. Like all his predecessors, he was mainly interested in solving the mystery of the ice ages.

Callendar's publications attracted some attention, and climatology textbooks of the 1940s and 1950s routinely included a brief reference to his studies. But most meteorologists gave Callendar's idea scant credence. In the first place, they doubted that CO_2 had increased at all in the atmosphere. The old data were untrustworthy, for measurements could vary with every change of wind that brought emissions from some factory or forest.Already in the nineteenth century scientists had observed that the level of the gas rose, for example, near a flock of sheep busy exhaling the gas, and dropped in London during the inactivity of a bank holiday.

If in fact CO_2 was rising, that could only be detected by a meticulous programme stretching decades into the future. The objections that had been raised against Arrhenius also had to be faced. Wouldn't the immense volume of the oceans absorb all the extra CO_2? Callendar countered that the thin layer of ocean surface waters would quickly saturate, and it would take thousands of years for the rest of the oceans to turn over and be fully exposed to the air. But nobody knew the actual turnover rate, and it seemed that the oceans would have time to handle any extra gases. According to a well-known estimate published in 1924, even without ocean absorption it would take 500 years for fuel combustion to double the amount of CO_2 in the atmosphere.

There was also the old objection, which most scientists continued to find decisive, that the overlapping absorption bands of CO_2 and water vapour already blocked all the radiation that those molecules were capable of blocking. Callendar tried to explain that the laboratory spectral measurements were woefully incomplete. Gathering scattered observational data, he argued that there were parts of the spectrum where the CO_2 bands did not overlap others. Some scientists found this convincing, or at least kept an open mind on the question. But it remained the standard view that, as an official U.S. Weather Bureau publication put it, the masking of CO_2 absorption by water vapour was a "fatal blow" to the CO_2 theory. Therefore, said this authority, "no probable increase in atmospheric CO_2 could materially affect" the balance of radiation. Most damaging of all, Callendar's calculations of the greenhouse effect temperature rise ignored much of the real world's physics. For example, as one critic pointed out immediately, he only calculated how heat would be shuttled through the atmosphere by radiation, ignoring the crucial energy transport by convection as heated air rose from the surface (this deficiency would haunt greenhouse calculations through the next quarter-century). Worse, any rise in temperature would allow the air to hold more moisture, which would probably mean more clouds.

Callendar admitted that the actual climate change would depend on interactions involving changes of cloud cover and other processes that no scientist of the time could reliably calculate. Few thought it worthwhile to speculate about such dubious questions, where data were rudimentary and theory was no more than hand-waving. Better to rest with the widespread conviction that the atmosphere was a stable, automatically self-regulated system. The notion that humanity could permanently change global climate was implausible on the face of it, hardly worth a scientist's attention.

The scientists who brushed aside Callendar's claims were reasoning well enough. (Subsequent work has shown that the temperature rise up to 1940 was, as his critics thought, mainly caused by some kind of natural cyclical effect, not by the still relatively low CO_2 emissions. And the physics of radiation and climate was indeed too poorly known at that time to show whether adding more gas could make much difference.) Yet if Callendar was mistaken when he insisted he could prove global warming had arrived, it was a fortunate mistake.

Research by definition is done at the frontier of ignorance. Like nearly everyone described in these essays, Callendar had to use intuition as well as logic to draw any conclusions at all from the murky data and theories at his disposal. Like nearly everyone, he argued for conclusions that mingled the true with the false, leaving it to later workers to peel away the bad parts. While he could not prove that global warming was underway, he had given reasons to reconsider the question. We owe much to Callendar's courage. His claims rescued the idea of global warming from obscurity and thrust it into the marketplace of scientific ideas. Not everyone dismissed his claims. Their very uncertainty attracted scientific curiosity.

The Speculations Vindicated

The complacent view that CO_2 from human activity could never become a problem was overturned during the 1950s by a series of costly observations. This was a consequence of the Second World War and the Cold War, which brought a new urgency to many fields of research. American scientists enjoyed massively increased government funding, notably from military agencies.

The officials were not aiming to answer academic questions about future climates, but to provide for pressing military needs. Almost anything that happened in the atmosphere and oceans could be important for national security. Among the first products were new data for the absorption of infrared radiation, a topic of more interest to weapons engineers than meteorologists.

The early experiments that sent radiation through gases in a tube, measuring bands of the spectrum at sea-level pressure and temperature, had been misleading. The bands seen at sea level were actually made up of overlapping spectral lines, which in the primitive early instruments had been smeared out into broad bands. Improved physics theory and precise laboratory measurements in the 1940s and after encouraged a new way of looking at the absorption. Scientists were especially struck to find that at low pressure and temperature, each band resolved into a cluster of sharply defined lines, like a picket fence, with gaps between the lines where radiation would get through. The most important CO_2 absorption lines did not lie exactly on top of water vapour lines. Instead of two overlapping bands, there were two sets of narrow lines with spaces for radiation to slip through.

So even if water vapour in the lower layers of the atmosphere did entirely block any radiation that could have been absorbed by CO_2, that would not keep the gas from making a difference in the rarified and frigid upper layers. Those layers held very little water vapour anyway. And scientists were coming to see that you couldn't just calculate absorption for radiation passing through the atmosphere as a whole, you had to understand what happened in each layer—which was far harder to calculate.

Digital computers were now at hand for such calculations. The theoretical physicist Lewis D. Kaplan decided it was worth taking some time away from what seemed like more important matters to grind through extensive numerical computations. In 1952, he showed that in the upper atmosphere, adding more CO_2 must change the balance of radiation significantly.

But would adding carbon dioxide in the upper layers of the air significantly change the surface temperature? Only detailed computations, point by point across the infrared spectrum and layer by layer up through the atmosphere, could answer that question. By 1956, such computations could be carried out thanks to the increasing power of digital computers. The physicist Gilbert N. Plass took up the challenge of calculating the transmission of radiation through the atmosphere, nailing down the likelihood that adding more CO_2 would increase the interference with infrared radiation. Going beyond this qualitative result, Plass calculated that doubling the level would bring a 3-4°C rise. Assuming that emissions would continue at the current rate, he expected that human activity would raise the average global temperature "at the rate of 1.1 degree C per century."

The computation, like Callendar's, paid no attention to possible changes in water vapour and clouds, and overall was too crude to convince scientists. "It is almost certain," one authority scolded, "that these figures will be subject to many strong revisions." Yet Plass had proved one central point: it was a mistake to dismiss the greenhouse effect with spectroscopic arguments. He warned that climate change could be "a serious problem to future generations"—although not for several centuries. Following the usual pattern, Plass was mainly interested in the way variations in CO_2 might solve the mystery of the ice ages. "If at the end of this century the average temperature has continued to rise," he wrote, then it would be "firmly established" that CO_2 could cause climate change

None of this work met the argument that the oceans would promptly absorb nearly all the CO2 humanity might emit. Plass had estimated that gas added to the atmosphere would stay there for a thousand years. Equally plausible estimates suggested that the surface waters of the oceans would absorb it in a matter of days. Fortunately, scientists could now track the movements of carbon with a new tool—the radioactive isotope carbon-14. This isotope is created by cosmic rays in the upper atmosphere and then decays over millennia. The carbon in ancient coal and oil is so old that it entirely lacks the radioactive isotope. In 1955, the chemist Hans Suess reported that he had detected this fossil carbon in the atmosphere.

The amount that Suess measured in the atmosphere was barely one per cent, a fraction so low that he concluded that the oceans were indeed taking up most of the carbon that came from burning fossil fuels. A decade would pass before he reported more accurate studies, which showed a far higher fraction of fossil carbon. Yet already in 1955 it was evident that Suess's data were preliminary and insecure. The important thing he had demonstrated was that fossil carbon really was showing up in the atmosphere. More work on carbon-14 should tell just what was happening to the fossil carbon. Suess took up the problem in collaboration with Roger Revelle at the Scripps Institution of Oceanography. (Some other carbon-14 experts attacked the topic independently, all reaching much the same conclusions.) From measurements of how much of the isotope was found in the air and how much in sea water, they calculated the movements of CO_2. It turned out that the ocean surface waters took up a typical molecule of CO_2 from the atmosphere within a decade or so. Radiocarbon data also showed that the oceans turned over completely in several hundred years, an estimate soon confirmed by evidence from other studies. At first sight that seemed fast enough to sweep any extra CO_2 into the depths.

But Revelle had been studying the chemistry of the oceans through his entire career, and he knew that the seas are not just salt water but a complex stew of chemicals. These chemicals create a peculiar buffering mechanism that stabilizes the acidity of sea water. The mechanism had been known for decades, but Revelle now realized that it would prevent the water from retaining all the extra CO_2 it took up. A careful look showed that the surface layer could not really absorb much gas—barely one-tenth the amount a naïve calculation would have predicted.

Revelle did not at first recognize the full significance of his work. He made a calculation in which he assumed that industry would emit CO_2 at a constant rate (like most people at the time, he scarcely grasped how explosively population and industry were rising). This gave a prediction that the concentration in the air would level off after a few centuries, with an increase of no more than 40%.

Revelle did note that greenhouse effect warming "may become significant during future decades if industrial fuel combustion continues to rise exponentially." He also wrote that "Human beings are now carrying out a large scale geophysical experiment of a kind that could not have happened in the past nor be reproduced in the future." As sometimes happens with landmark scientific papers, written in haste while understanding just begins to dawn, Revelle's explanation was hard to grasp. Other scientists failed to see the point that was obscurely buried in the calculations, and continued to deny there was a greenhouse effect problem. In 1958, when Callendar published a paper to insist once again that CO_2 observations showed a steady rise from the 19th century, he noted Revelle's paper but still confessed that he did not understand why "the oceans have not been accepting additional CO_2 on anything like the accepted scale."

Finally in 1959 two meteorologists in Sweden, Bert Bolin and Erik Eriksson, caught on. They explained the sea water buffering clearly—so clearly that during the next few years, some scientists cited Bolin and Eriksson's paper for this decisive insight rather than Revelle and

Suess's (only in later years was Revelle always cited for the discovery). The central insight was that although sea water did rapidly absorb CO_2, most of the added gas would promptly evaporate back into the air before the slow oceanic circulation swept it into the abyss.

To be sure, the chemistry of air and sea water would eventually reach an equilibrium—but that could take thousands of years. Arrhenius had not concerned himself with timescales shorter than that, but geoscientists in the 1950s did. In the late 1950s a few American scientists, starting with Plass, tentatively began to inform the public that greenhouse gases might become a problem within the next few centuries. Revelle in particular warned journalists and government officials that greenhouse warming might come within the foreseeable future, and deserved serious attention.

The stakes were revealed when Bolin and Eriksson pursued the consequences of their calculation to the end. They assumed industrial production would climb exponentially, and figured that atmospheric CO_2 would rise some 25% by the year 2000. That was a far swifter rise than anyone before had suggested. As the New York Times reported in a brief note, Bolin suggested that the effect on climate "might be radical." In 1962, a still stronger (although also little heeded) warning was sounded by the Russian climate expert Mikhail Budyko. His calculations of the exponential growth of industrial civilization suggested a drastic global warming within the next century or so.

Once meteorologists understood that ocean uptake was slow, they found it possible that CO_2 levels had been rising, just as Callendar insisted. Yet it was only a possibility, for the measurements were all dubious. By the mid 1950s, researchers were saying that it was important to measure, much more accurately, the concentration of CO_2 in the atmosphere. A Scandinavian group accordingly set up a network of 15 measuring stations in their countries. Their only finding, however, was a high noise level. Their measurements apparently fluctuated from day to day as different air masses passed through, with differences between stations as high as a factor of two. Only much later was it recognized that their methods of analyzing the air had been inadequate, and responsible for much of the noise.

A leading authority summarized the scientific opinion of the late 1950s: "it seems almost hopeless to arrive at reliable estimates [of CO_2]... by such measurements in limited areas." To find if the gas level was changing, measurements would have to "be made concurrently and during a great number of years" at many locations. Charles David (Dave) Keeling held a different view. As he pursued local measurements of the gas in California, he saw that it might be possible to hunt down and remove the sources of noise. Technical advances in infrared instrumentation allowed an order of magnitude improvement over previous techniques for measuring gases like CO_2. Taking advantage of that, however, would require many costly and exceedingly meticulous measurements, carried out someplace far from disturbances. Most scientists, looking at the large and apparently unavoidable fluctuations in the raw data, thought such precision irrelevant and the instrumentation too expensive. But Revelle and Suess had enough funds, provided by the International Geophysical Year, to hire Keeling to measure CO_2 with precision around the world.

Revelle's simple aim was to establish a baseline "snapshot" of CO_2 values around the world, averaging over the large variations he expected to see from place to place and from time to time. After a couple of decades, somebody could come back, take another snapshot, and see if the average CO_2 concentration had risen. Keeling did much better than that with his new instruments.

With painstaking series of measurements in the pristine air of Antarctica and high atop the Mauna Loa volcano in Hawaii, he nailed down precisely a stable baseline level of CO_2 in the atmosphere. In 1960, with only two full years of Antarctic data in hand, Keeling reported that this baseline level had risen. The rate of the rise was approximately what would be expected if the oceans were not swallowing up most industrial emissions.

Lack of funds soon closed down the Antarctic station, but Keeling managed to keep the Mauna Loa measurements going with only a short hiatus. As the CO_2 record extended it became increasingly impressive, each year noticeably higher. Soon Keeling's curve, jagged but inexorably rising, was widely cited by scientific review panels and science journalists. For both scientists and the public it became the primary icon of the greenhouse effect.

Index

S

T

V

W

Y